FROM HIROSHIMA TO HARRISBURG:

The Unholy Alliance

Inside front and back cover: Views of devastated Hiroshima (photos – front: Illustrated London News; back: Keystone Press Agency)

FROM HIROSHIMA TO HARRISBURG:

The Unholy Alliance

JIM GARRISON

SCM PRESS LTD

334 00504 3

First published 1980
by SCM Press Ltd
58 Bloomsbury Street, London

Filmset by C. Leggett & Son Ltd
and printed in Great Britain by
Richard Clay Ltd (The Chaucer Press)
Bungay, Suffolk

Dedicated to
Danny Paul Nelson and Kimberly Anne Drew,
recently born, and their generation.

Dedicated also to
Frank Thomas, a Commanche and Senecca medicine man
who was killed by those who do not understand
on 14 May 1979.
He began the fulfilment of the ancient Indian prophecies
concerning the rainbow warriors
by empowering all who came to him with the awareness
that non-violence is born of a reverence for the earth.
For the rainbow warrier, Frank lives.

One early dawn . . . a dazzling flash of light, strangely brilliant in quality, illumined the most distant peaks, eclipsing the first rays of the rising sun. There followed a prodigious burst of sound . . . The thing had happened. For the first time on earth an atomic fire had burned for the space of a second, industriously kindled by the science of Man.

But having thus realised his dream of creating a new thunder-clap, Man, stunned by his success, looked inward and sought by the glare of the lightning his own hand had loosed to understand its effect upon himself . . . It is no longer a question of laying hands upon existing forces freely available for his use. This time a door has been decidedly forced open, giving access to a new and supposedly inviolable compartment of the universe. Hitherto Man was using matter to serve his needs. Now he has succeeded in seizing and manipulating the sources commanding the very origins of matter — springs so deep that he can release for his own purposes what seemed to be the exclusive property of the sidereal powers, and so powerful that he must think twice before committing some act which might destroy the earth.

Teilhard de Chardin,
'Some Reflections on the Spiritual Repercussions of the Atom Bomb', *The Future of Man*.

CONTENTS

INTRODUCTION

Arthur Koestler recently remarked that if he was to choose the most important date in all of human history he would without hesitation choose 6 August 1945, the day the United States dropped an atomic bomb on the Japanese city of Hiroshima. Before that date, he said, each person had to live with only his or her own individual death in mind: but since Hiroshima, we have all been forced to live with the prospect of the extinction of our entire species.

What occurred on the day, according to Albert Einstein, was a quantum leap in technology that changed everything in the world – except for human consciousness. Not comprehending our leap in technology and how it had dramatically altered the ways of peace and the methods of war, we have continued to arm ourselves with weapons of mass destruction to the point of thrusting history into the Age of Overkill, threatening each nation, each city, each individual with death not just once but scores of times over. World peace is insured not by understanding among the peoples but through a balance of terror between the nations.

What makes the situation critical is that nuclear weapons are proliferating to countries beyond the control of the superpowers – to countries such as Israel, India, South Africa, Brazil, Argentina, and Pakistan. The balance of terror has increasingly become a spiral of competition between rival groups of nations jostling each other for control over their part of the globe. As nuclear weapons are the ultimate symbol of power, all seek them, paradoxically fashioning their atoms for war from the plutonium waste produced from their atoms for peace programmes. Integral to the proliferation of nuclear weapons is the presence and proliferation of nuclear reactors.

What is increasingly being recognized is that while sold under the slogan 'atoms for peace', nuclear reactors are a major danger to human life and the environment in their own right. The incident at Harrisburg that forced thousands of people to evacuate their homes and came close to irradiating a large section of the eastern coast of the United States symbolizes – as Hiroshima does for nuclear weapons – the dangers of the plutonium economy.

The history of the forty years between the inception in 1939 of the secret programme to build the atomic bomb and the near meltdown of the nuclear reactor at Harrisburg in 1979 is the subject of this book. It is neither an economic analysis nor a work that emphasizes the scientifically technical aspects of nuclear energy: rather, it stresses the *human* dimension of nuclear history, attempting to discern the impact of the nuclear weapons/nuclear reactor complex upon our persons – upon our psyches, upon our bodies, upon our freedoms. Because this history emphasizes the effects of nuclear power on people, the life and death of one woman, Karen Silkwood, will be offered towards the conclusion as a symbol not only of what the plutonium economy is capable but also of what each individual can do to counter it.

Fundamentally, the problem of nuclear power is seen in terms of Einstein's assessment, that a leap in consciousness – in human self-awareness and morality – is needed if it is to be dealt with adequately. Because in splitting the atom we made a leap in technology *without* making this parallel leap in consciousness, Einstein warned that we are headed toward 'unparalleled catastrophe'. We have become as it were stone age minds with atomic age weapons.

I have written this book to help to make the necessary leap in consciousness in order to avoid catastrophe.

I

Hiroshima and the Advent of the Atomic Age

A. THE MANHATTAN ENGINEER DISTRICT PROJECT

The last scientific visitors to Berlin in 1939 reported that work was being carried on in Germany on the separation of uranium 235. News and rumours subsequent to the beginning of the Second World War indicated further Nazi research into both the thermal diffusion method of isotope separation and in heavy water, a possible moderator in a nuclear reactor capable of producing plutonium.[1] Scientists knew that plutonium was the essential ingredient in the construction of a fission bomb, and fear became widespread that Hitler might develop an atomic bomb for use against the Allies. For this reason, scientists in Britain and the United States felt compelled to develop a fission bomb as well, believing that such a weapon could not be allowed to be developed by the Nazis first.[2]

So great was the fear of a Nazi lead in this critical area of atomic fission that on 2 August 1939, after strong encouragement from the Hungarian physicist Leo Szilard, Albert Einstein wrote a letter at his holiday home at Peconic, Long Island, warning President Roosevelt that research into nuclear fission in Germany and elsewhere could lead to a major scientific breakthrough with direct implications for warfare: '. . . extremely powerful bombs of a new type may thus be constructed,' he said, and he advised Roosevelt to begin American research and development into this area. A friend of Einstein's, financier Alexander Sachs, agreed to take the letter to the White House, although it was not until 11 October 1939 that he obtained an audience with the President. After hearing Einstein's counsel, Roosevelt dismissed Sachs, stating that for the US to become involved at this stage would be 'premature'.

Sachs was mortified by the President's apparent inability to grasp what Einstein was attempting to convey and asked for another meeting. Roosevelt agreed to a breakfast meeting the next day. As Sachs recalls the meeting, Roosevelt was again unimpressed until in exasperation Sachs told him the story of Robert Fulton, who attempted to sell his newly invented steamship to Napoleon. 'Mr President,' said Sachs, 'he took it to Napoleon who said it was impractical. Napoleon thus lost for France the vessel which would have allowed him to invade England and claim victory.'

At this remark the President grew serious and after some moments of silence finally asked. 'What you are after is to see that the Nazis don't blow us up?'

'Exactly.'

Roosevelt then summoned his military aide, General Edwin 'Pa' Watson, and, handing him the Einstein letter along with the supporting documents from Szilard, said, 'Pa, I want action on this.'[3]

These words began the project that was to culminate some four and a half years later in the atomic bombing of Hiroshima and Nagasaki. According to Roosevelt's Secretary of War, Henry L. Stimson, the President's policy was a simple one: 'It was to spare no effort in securing the earliest possible successful development of an atomic weapon.'[4]

British scientists were also researching the possibilities of a nuclear device by this point, and by the spring of 1941 were certain that one could be made. Throughout 1941 and 1942, however, it was believed by the Allies that the Nazi researchers had the lead. Even as late as 1943, Roosevelt believed this to be the case, particularly after Niels Bohr made his escape from Denmark and reported that the eminent German physicist Werner Heisenberg was directing atomic research for the Nazis.[5]

Enormous questions needed to be answered and experiments done before any real thought could be given to actually designing an experiment in criticality which would enable scientists to know whether the entire concept of a bomb was feasible. By August 1942, the Manhattan Engineer District Project was organized, centred at the University of Chicago, to conduct the final experiments.[6] The Scientific Director of the project was Julius Oppenheimer; the Military Commander, Major General Leslie R. Groves. On 2 December 1942, Enrico Fermi, who directed the chain reaction experiments in Chicago, finally produced the chain reaction necessary actually to detonate the bomb. The pressures upon the scientists had been immense, for there had been speculation that if the chain reaction could not be controlled, Chicago rather than an enemy city would be the first to experience an atomic blast.

The scientists involved in the Chicago experiments termed the chain reaction process the 'K Factor', calling it among themselves 'the great god K'.[7] Having succeeded in creating 'K', however, the next question presented itself: how to enclose 'K' inside a bomb casing. Again large technical problems had to be investigated and solved, but by 1943, scientists were certain of success. Throughout this period, Roosevelt was authorizing vast sums of money for the project out of 'hidden' accounts in the Federal budget.[8] He had also managed to keep the entire programme top secret, even from certain members of his own cabinet and from the members of Congress. Even the one person who above all should have been informed, Vice President Truman, was kept completely in the dark, a misjudgment on the part of Roosevelt which was to prove to be, according to Edward Teller, 'a terrible and unforgivable mistake'.[9]

On 12 April 1945, Roosevelt died while resting in Warm Springs, Georgia. Truman, still ignorant of the Manhattan Project, was sworn in as President that evening. His first act was to call a cabinet meeting, after which Stimson remained behind, telling Truman that he must speak to him concerning 'a most urgent matter'. He then informed the new President that an immense enterprise was under way, 'a project looking to the development of a new explosive of almost unbelievable destructive power'.[10] Stimson did not elaborate, however, leaving that to the Director of War Mobilization, James F. Byrnes, who the next day informed the President that the US was 'perfecting an explosive great enough to destroy the whole world'.[11]

By this time, Stimson particularly was becoming increasingly caught up with the accelerating pace of the atomic bomb research and production. Next to the President, ultimate authority was his, and so he began spending an increasing amount of time reflecting upon the political and long term implications of its use. However, while Roosevelt had 'understood the terrible responsibility involved in our attempt to unlock the doors to such a devastating weapon', Stimson, recalling in his memoirs that Roosevelt spoke to him many times of the 'catastrophic potentialities' of the atomic research being done,[12] was unsure as to whether Truman felt the gravity of the situation in the same way.[13] Rather, Truman seemed to view the weapon as little more than a super-bomb to be used for tactical military reasons much like any other weapon. Stimson prepared a memorandum, therefore, dealing not so much with the military use of the bomb as with its long-term political meaning. It emphasized the fact that 'within four months we shall in all probability have completed the most terrible weapon ever known in human history, one bomb of which could destroy a whole city'. The memo went on to say that although the US shared with Britain the only imminently successful atomic bomb research and development 'at present', it was practically certain that 'we could not remain in this position indefinitely', specifying Russia as 'the only nation that could enter into production within the next few years'. International control of some sort was imperative, therefore, because 'modern civilization might be completely destroyed' if atomic bomb production went unchecked. Stimson was gravely worried about this possibility because the world 'in its present state of moral advancement compared with its technical development would be eventually at the mercy of such a weapon', if controls were not solidified. The memo concluded by recognizing that 'no system of control heretofore considered would be adequate to control this menace', although it was a question that had to be answered. Stimson stated that he felt that the question of control and sharing the bomb with others was the 'primary question of our foreign relations'.[14]

Stimson met the President along with Major General Groves on 25 April and read the entire memo out loud to him. Truman complained several times about the memo's length but Stimson and Groves insisted that it was as concise as possible. After Truman indicated general agreement with the substance of what was read, Stimson then proposed that a committee be established and 'charged with the function of advising the President on the various questions raised by our apparently imminent success in developing an atomic weapon'.[15]

Truman approved of the concept, and on 8 May a select group of high administration officials and several university presidents met at the Pentagon. Stimson as chairperson spoke first and the recording secretary states that he 'made it very plain that he thought the members of the committee should concern themselves not only with the military application but with any and all questions that would be raised by the advent of this new force'. It was crucial that all involved should understand, said Stimson, that 'this development really represented a new relationship between man and the universe', and their work 'should be conducted in line with that fact'. This meant specifically that the Interim Committee (as the group was called) was to address itself to 'publicity, to the questions of an open society versus a closed society, the matter of international control, what would be our relations with our allies, particularly how should we handle this question *vis-à-vis* the Soviet Union, what should the postwar organization be for fostering research in this field at home and abroad'.[16]

Of particular note in the Interim Committee was the Scientific Panel formed to advise it. On it were four eminent scientists: Oppenheimer, now Director of the Los Alamos Laboratory; Arthur H. Compton of the Metallurgical Laboratory of Chicago; Ernest O. Lawrence, Director of the Berkeley Radiation Laboratory; and Enrico Fermi.[17]

On 1 June after less than three weeks of deliberations, the Interim Committee, in consultation with the Scientific Panel, made the following recommendations: first, that the bomb be dropped upon Japan as soon as possible; second, that it be used against a dual target, meaning a military installation surrounded by a civilian community; and third, that the bomb be dropped without prior warning.[18] The larger issues which Stimson had charged the committee to discuss were never thoroughly aired; instead the committee, like Truman, focussed its attention on the tactical questions involved in compelling an unconditional surrender on the part of the Japanese. Indeed, Stimson states specifically that the factor uppermost in all of their minds was that 'the principal political, social, and military objective of the United States in the summer of 1945 was the prompt and complete surrender of Japan'. The committee was convinced, he said, that 'only the complete destruction of (Japan's) military power could open the way to lasting peace'.[19]

B. THE MILIEU OF WAR

In the light of the present situation over nuclear weapons it is difficult to understand the short-sightedness of the decision-makers involved in determining the fate of the atomic bomb.

Nevertheless, it is necessary to be sensitive to the milieu of war in which they were operating, a war that had been dragging on for over four years and which everyone was eager to finish. That they would spend less than three weeks to make what has turned out to be a most momentous decision can be attributed in part to the fact that at the time, bringing Japan to her knees was not thought to be an easy matter. Indeed, the intelligence section of the War Department General Staff estimated Japanese military strength at over 5,000,000 soldiers.[20] While the Japanese navy had been all but decimated except as a harrying force against the invading fleets, and their air force had been reduced to reliance primarily upon kamikaze attacks, the army was still largely intact and fully capable of mounting a sustained defence of the Japanese mainland. In short, while certain of victory, at the time the Allies expected a long and costly campaign ahead.

The strategic plans for the Allied armed forces for the defeat of Japan had been prepared without reliance upon the atomic bomb. They were centred upon Operation OLYMPIC, which called for an intensified sea and air blockade throughout the summer to be followed on 1 November by a land invasion of the southern Japanese island of Kyushu. This would be followed in turn by the invasion of the main island of Honshu in the spring of 1946.[21] The number of American soldiers expected to be involved in this operation numbered over 5,000,000. The Joint Chiefs of Staff expected the invasion to last as long as one and a half years and had informed Stimson to expect over a million casualties.[22] General Douglas MacArthur, Supreme Commander of US forces in the Pacific, reckoned on 50,000 US casualties in the first thirty days of the Kyushu invasion alone.[23] Churchill, who planned for the British to fight alongside the Americans, recalls in his account of the Second World War that the number of expected casualties were put as high as they were because 'we had contemplated the desperate resistance of the Japanese fighting to the death with Samurai devotion, not only in pitched battles, but in every cave and dug-out'. He also estimated a million US casualties and added to that 500,000 British casualties.[24] Truman put the minimum number of expected US casualties at 500,000, basing this estimate on the assumption that the Soviets would enter the war against Japan and keep at least 2,000,000 of the Japanese forces engaged on the Chinese mainland.[25]

The spectre of an invasion was particularly agonizing for the

Americans, given the kamikaze mentality of the Japanese fighters. During the Okinawa campaign alone, sixteen US ships had been sunk and 185 damaged by kamikaze action. By July 1945, the total sunk by the kamikaze attacks was thirty-four ships, including three aircraft carriers; 285 were damaged.[26] In preparation for the expected US invasion, the US Strategic Bombing Survey estimated that the Japanese Air Force had readied 5,350 suicide planes.[27] The Japanese Navy was also preparing suicide boats. According to Captain Tsuezo Wachi, commander of a unit of these 'shimyo boats' for the defence of southern Japan, 600 such 'human torpedoes' had been readied in the area of his command alone, capable of sinking 200 US transport ships.[28]

The world-view within which the American policy makers were operating has been most succinctly summarized by Stimson, who perhaps more than anyone concerned himself with the political and long-term implications of using the atomic bomb. In an article he wrote shortly after the war was over, entitled 'The Decision to Use the Atomic Bomb', he states that uppermost in all of their minds was the fact that

> . . . two great nations were approaching contact in a fight to a finish which would begin on November 1, 1945. Our enemy, Japan, commanded forces of somewhat over 5,000,000 armed men. Men of these armies had already inflicted upon us, in our breakthrough of the outer perimeter of their defences, over 300,000 battle casualties. Enemy armies still unbeaten had the strength to cost us a million more. . . .
>
> In light of the formidable problem which thus confronted us, I felt that every possible step should be taken to compel a surrender of . . . all Japanese troops. . . .[29]

The policy adopted by President Truman and the Allies, according to Stimson, was that 'the only road to early victory was to exert maximum force with maximum speed . . .' and that within this policy '. . . if victory could be speeded by using the bomb, it should be used; if victory must be delayed in order to use the bomb, it should *not* be used'.[30]

In Stimson's mind, the US held 'two cards' to assist it in attempting to bring Japan to the point of unconditional surrender in the shortest possible time: one was the traditional veneration given to the Emperor by his subjects and the other was the atomic bomb, which Stimson felt should be used in a manner 'best calculated to persuade the Emperor and the counsellors about him to submit. . . .' This is an important point to note because it means, as Stimson himself observes, that 'the atomic bomb was more than a weapon of terrible destruction; it was a *psychological* weapon'.[31]

Stimson argues that, as a psychological weapon, 'the bomb . . . served exactly the purpose we intended'. This meant specifically that

> the peace party was able to take the path of surrender, and the whole weight of the Emperor's prestige was exerted in favour of peace. When the Emperor ordered surrender, and the small but dangerous group of fanatics who opposed him were brought under control, the Japanese became so subdued that the great undertaking of occupation and disarmament was completed with unprecedented ease.[32]

As for the effectiveness of this policy, Stimson points out that:

> In March, 1945, our Air Forces had launched the first great incendiary raid on the Tokyo area. In this raid more damage was done and more casualties were inflicted than was the case at Hiroshima . . . but the Japanese fought on. On August 6, one B-29 dropped a single atomic bomb on Hiroshima. Three days later a second bomb was dropped on Nagasaki, and the war was over.[33]

As Churchill was to put the matter in his account of the Second World War, the nightmare of prolonged fighting with hundreds of thousands of casualties vanished with the atomic bomb. 'In its place,' he says, 'was the vision – fair and bright indeed it seemed – of the end of the whole war in one or two violent strokes.'[34]

While the *participants* in the decision making by and large concur with Stimson's and Churchill's official statements concerning the reasoning that went into dropping the atomic bomb on Japan, many *commentators* argue that their rationale is oversimplistic, if not patently untrue. Most notable in this category are Gar Alperovitz, Paul Kecskemeti and Herbert Feis, all of whom cite the United States Strategic Bombing Survey, done after touring Japan at the conclusion of hostilities. The survey concluded that

> certainly prior to 31 December 1945, and in all probability

> prior to 1 November 1945, Japan would have surrendered even if the atomic bombs had not been dropped, even if Russia had not entered the war, and even if no invasion had been planned or contemplated.[35]

Indeed, the fire-bombing of Japanese cities had been devastating, particularly the incendiary bombing of Tokyo on 9 March 1945, which completely burned out 15.8 square miles, destroying 267,171 buildings and killing 83,793 Japanese.[36] By the end of March, Air Force General Curtis LeMay had fire-bombed Osaka, Kobe and Nagoya twice, completely destroying over thirty-two square miles of these cities, all with less than a 1% casualty rate in the US bombing crews. The joint chiefs then designated thirty-three other urban areas for similar devastation, and by the end of April LeMay was flying over 3,000 sorties to various parts of Japan, estimating the number to grow to 7,000 by September. After that, he wrote back in a memo to the joint chiefs that he would have to quit 'as we were running out of targets. . . .'[37] Giovannitti and Freed assert with Robert Butow that as a result of this massive aerial bombardment during the spring of 1945, 'Japan was literally losing the war faster than the United States and her allies were winning it'.[38]

Alperovitz argues that this awareness was known inside the US military and cites General Eisenhower's reaction when Stimson informed him of the atomic bomb at Potsdam, telling Eisenhower that it was to be used against Japan:

> During his recitation of the relevant facts, I had been conscious of feeling a depression and so I voiced to him my grave misgivings, first on the basis of my belief that Japan was already defeated and that dropping the bomb was completely unnecessary, and secondly because I thought that our country should avoid shocking world opinion by the use of a weapon whose employment was, I thought, no longer mandatory as a measure to save American lives. It was my belief that Japan was, at that very moment, seeking some way to surrender with a minimum loss of 'face'.[39]

Hewlett and Anderson point out that MacArthur, who like Eisenhower was only *informed* of the decision to use the bomb rather than brought into the decision-making process itself, had similar reservations.[40]

Kecskemeti amplifies this point by stressing that saving face for the Japanese by this point meant merely saving the institution of the Emperor. He argues that 'the Japanese would have surrendered without being bombed, if the United States had clearly declared its readiness to retain the Emperor'.[41] In fact, Kecskemeti goes so far as to state that it was the American willingness to spare the Emperor which 'alone' induced Japan to surrender. Feis maintains here that by clarifying this critical point earlier in the American demands for unconditional surrender, the war could have conceivably ended much sooner than it did.[42]

Kecskemeti raises another important point, namely that 'the main factor that determined the timing of the surrender note was the Soviet declaration of war', *not* the dropping of the atomic bombs.[43] Alperovitz concurs with this and recalls General Marshall's appraisal of 18 June 1945, that 'the impact of Russian entry on the already hopeless Japanese may well be the decisive action levering them into capitulation'.[44] While acknowledging, therefore, the claim made by Truman that the bomb 'saved millions of lives',[45] Alperovitz states quite categorically that the facts of the matter were that *'before the atomic bomb was dropped each of the Joint Chiefs of Staff advised that it was highly likely that Japan could be forced to surrender "unconditionally", without use of the bomb and without an invasion'*.[46] Their reasoning for their counsel followed the appraisal of General Marshall. Hewlett and Anderson amplify this by observing that even Churchill believed that it was Russia's entry into the war rather than the use of the atomic bombs that compelled the Japanese to surrender: 'It would be a mistake to suppose,' he said, 'that the fate of Japan was settled by the atomic bomb. Her defeat was certain before the first bomb fell. . . .'[47]

The commentators, then, raise three points to counter the official position that the atomic bombs were a necessary method of ending the war in the Pacific: first, that Japan was already defeated industrially; second, that Japan was willing to surrender given an assurance that the institution of the Emperor be allowed to continue; and third, that it was the Russian entry into the war rather than the bombs that compelled the Japanese surrender.

While certainly not wanting to defend the use of the atomic bomb in any way, I believe these criticisms are lacking in credibility. While the hindsight of such reports such as that of the US Strategic Bombing Survey indicate that Japan would have collapsed before the year was out without either atomic bomb, the Russian entry into the war or the Allied invasion, because of its evaluation of the industrial destruction inflicted upon the Japanese urban centres, all evidence points to the fact that the will of the people and the military was strong right up until the combined dropping of the atomic bombs and the Russian entry into the war against them. The assumption that industrial devastation can be readily translated into political surrender is not borne out by historical fact. The Germans in the European theatre of the war held out far longer than the Allies thought they could. Perhaps the most recent example has been that of North Vietnam during the Indo-China War. It was able not only to absorb a virtually complete decimation of its industrial sector because of aerial bombings – the US dropped more bombs on Vietnam than all the bombs dropped by all sides during the Second World War – but also to go forward and actually defeat the American forces. In war, what determines the surrender point is not material destruction but the breaking of will. By the spring of 1945, this was something the Allies clearly had not done to Japan, despite the fact that they had pulverized her major urban centres. By April, even though the Japanese military recognized that the war was unwinnable, the situation was such that, according to the Chief Cabinet Secretary of the Japanese Supreme Council, Hisatsune Sakomizu, 'The Army was determined to continue the war. Therefore if anyone had said "end the war" he might have been arrested by the military police. Things were like that then.'[48] Even such a man as Prime Minister Suzuki, who privately wished to surrender, made the pledge on 7 April that Japan's 'only course' would be to 'fight to the end'.[49] The military believed that with a combination of suicide ships and planes along with the rest of its naval and air forces and still formidable ground strength Japan could repulse the American invasion and in any case make that invasion so costly that a negotiated peace would ensue.

Psychologically, therefore, the nation still had cohesion and a sense of inner strength and purpose. Those seeking peace, such as the Emperor and the Prime Minister, sought a *negotiated* peace which they hoped would be mediated by the Soviet Union. At no point prior to 10 August was even the Emperor prepared to capitulate to *unconditional surrender*, the American demand. And on this demand the Americans were adamant, perhaps because of their inclination for revenge after Pearl Harbour, perhaps because of the impulse to demonstrate total US supremacy in the Pacific as in Europe. The Germans had surrendered unconditionally and the Japanese were to do the same. It will, of course, never be known whether Japan would have considered unconditional surrender earlier had the Americans made it clear earlier that this unconditional surrender meant the continuance of the institution of the Emperor. What is clear, however, is that in deciding *not* to make this clarification, Truman was acting upon the advice of the joint chiefs, the Secretary of War and the Secretary of State. In a memo dated 9 June from Chief of Staff Marshall to Secretary Stimson, Marshall defended this position, stating 'We must be careful to avoid giving any impression that we are growing soft.'[50] Stimson in time changed his mind and sided with Under Secretary of State Grew in advising that this clarification be made, but the final ultimatum issued from Potsdam in late July left the clarification out, the reasoning being that any magnanimous gesture on the part of the United States should come after unconditional surrender, not before. It was believed that any gesture before could only serve to strengthen the Japanese will to resist.[51] While *militarily*, therefore, Japan was finished in terms of her ability to make war, *psychologically* she was still combative; it was this that the Americans were attempting to break, Stimson believing, and most other policy-makers with him, that as a psychological weapon 'the bomb . . . served exactly the purpose we intended'.

It is worth observing in this regard that while several people spoke out later against the use of the atomic bomb, no one *at the time* counselled in an open and direct way that it should not be dropped; no one, that is, except for a few scientists involved, something to be discussed shortly. Illustrative of this point is the attitude of the US Strategic Bombing Survey, which while asserting *after* the war that use of the bomb was unnecessary, was *at the time* of the decision directing attention to the fact that the Japanese had over 5,000 suicide planes prepared against the US invasion.

Nevertheless, I believe the commentators are essentially correct in maintaining that the military situation *vis-à-vis* Japan did not make *imperative* the use of the atomic bomb, particularly since the Soviet entry into the war seemed to be as much of a factor in Japan's psychological collapse as the bomb itself. If someone such as Feis is correct, then, in asserting that 'there cannot be a well-grounded dissent from the conclusion (that) the use of the bomb was not *essential*'[52] in bringing about the defeat of Japan, were there other factors, not having to do with Japan perhaps, that entered into the decision-making?

Alperovitz suggests that there were, asserting that the American use of the bomb against Japan was directed as much against the Soviets *politically* as it was against the Japanese *militarily*.[53] He points to the fact that in terms of the Pacific theatre of the war, the American policy-makers from at least mid-May onwards were endeavouring to end hostilities before Soviet troops entered Manchuria and began encroachments in northern Asia as they were doing in Eastern Europe. While recognizing that Russian entry into the war against Japan would insure the Japanese capitulation, this is what they were actively seeking to avoid, although publicly the Americans continued to call for the Soviets to join them in the war against Japan. A few 'violent strokes', then, is what the Americans were looking for to close the war, not only because they had no desire for a costly invasion of Japan itself but because they had no wish to have to share rule with the Russians in any occupation of Japan.

However, the atomic bomb had an impact on diplomacy even in the European theatre. Indeed, the major conclusion that can be drawn from a study of American policy in general during the months of April-August of the Truman administration is that the atomic bomb influenced in a profound way the perspective of American policy makers.[54] Representative of many is the statement of Admiral Leahy that 'One factor that was to change a lot of ideas, including my own, was the atom bomb. . . .'[55] As Alperovitz points out, however, the change did not produce any new policies, particularly with regard to the Soviet Union; rather, 'the atomic bomb *confirmed* American leaders in their judgment that they had sufficient power to affect developments in the border regions of the Soviet Union'.[56] Truman, e.g., was quite specific in stating that possessing the atomic bomb gave him 'an entirely new feeling of confidence'.[57]

This 'feeling of confidence' is of profound importance because it was to make a decisive impact on later developments in the post-war period. To understand this it is necessary to recall that before the atomic bomb, many Western policy makers harboured grave doubts about the ability of the Allies to challenge the Soviet encroachment into Eastern Europe, particularly since it was doubtful whether the American public would permit the retention of large numbers of conventional troops after the war.

Without troops in Europe in large enough numbers, only foreign aid and large credit were left as the concrete bargaining lever the Americans could wield.

Churchill's estimate of the situation in 1944 exemplifies the attitude held by Western European policy makers in their attempts to judge Western strength *vis-à-vis* Soviet strength. When Roosevelt rejected his plea for an invasion through the Balkans, Churchill immediately foresaw that Allied power in post-war South Eastern Europe would be minimal, threatening even the British position in Greece. He wrote to Roosevelt just prior to the Russian invasion of Rumania and said that the only way to prevent utter anarchy was through attempts at 'persuasion', saying that, 'considering that neither you nor we have any troops there at all . . . they will probably do what they like anyhow'.[58]

Without troops to back up demands, 'the arrangements made about the Balkans were, I was sure . . . the best possible', Churchill recalls.[59]

By the time of the Yalta Conference, Roosevelt had come to share Churchill's assessment.[60] At the conference, the chief concern of both American and British diplomats as regards Eastern Europe was, to quote C. E. Black, 'to negotiate a compromise which would acknowledge the basic Russian security requirements without formally recognizing a Russian sphere of influence'.[61] Truman's Balkan representative recalled later about the conference that it was 'fateful that these discussions should have been held at a time when Soviet bargaining power in Eastern Europe was so much stronger than that of the Western allies'.[62] Byrnes, in summarizing the relative strengths of the Western Allies *vis-à-vis* the Soviet Union in early 1945, stated that 'It was not a question of what we would *let* the Russians do, but what we could *get* them to do'.[63]

Such, then, was the prevailing attitude among American and British diplomats during the spring of 1945, when Roosevelt was negotiating at Yalta. With the certainty of success in detonating the atomic bomb, however, the American attitude began to change, Alperovitz suggesting Truman's hard line on the issue of Poland in the summer of 1945 as the point where the new weapon's impact began to be felt.[64] In all cases, by July, American policy makers, specifically and most emphatically Byrnes, were convinced that the atomic bomb would allow the US to take a 'firm' stand in subsequent negotiations. As Churchill stated the matter: 'We now had something in our hands which would redress the balance with the Russians.'[65] Byrnes was even more explicit, counselling Truman that 'the Bomb might well put us in a position to dictate our own terms. . . .'[66] Alperovitz argues that it was this factor of the atomic bomb which was catalytic in Truman's departure from Roosevelt's policy of co-operation and was what convinced him to launch 'a powerful foreign policy initiative aimed at reducing or eliminating Soviet influence from Europe'.[67]

Stimson essentially agreed with this position and argued against those who sought a showdown with the Russians in the spring, promising that if a confrontation could be postponed until the US had demonstrated the devastating impact of the bomb, the Western Allies would be possessive of a 'decisive influence on our relations with other countries'.[68] After the premature showdown over Poland in May, Truman accepted Stimson's policy outlines. The Potsdam meeting in particular illustrates how this Stimson policy to wait for the bomb dominated American thinking from mid-May until early August.[69] Once the detailed report of the Alamorgordo test arrived, even someone such as Churchill, who had argued strenuously for an 'early and speedy showdown' with the Russians was convinced of Stimson's advice. 'We were', he said, 'possessive of powers which were irresistible.'[70]

The Potsdam conference took place, therefore, with both sides believing time was on their side, Truman because of the bomb and Stalin because American troops were to be shortly withdrawn from Europe. This logic insured the final outcome – deadlock, and the beginning of what General de Gaulle perceptively saw as 'unlimited friction between Soviet and Anglo-American participants'.[71]

What emerges from an analysis of the situation of war in which the Interim Committee and the different American and British political and military leaders found themselves, then, were two considerations that dominated the discussions concerning the use of the atomic bomb: the first was the military desire to end the war with Japan using 'maximum force with maximum speed', preferably before the Soviets could enter and insist on the joint occupation of Japan with the US; the second was the desire on the part of the Americans to gain political leverage in Europe against the continuing encroachment of Russian troops in eastern Europe.

The military consequences were immediate and definite: within a week Japan had surrendered unconditionally. The political ramifications, however, were far less tangible, particularly with regard to the Soviets. As we shall see in the next section, little was done except to sow the seeds of distrust and hostility that have come to bloom in our day with the Age of Overkill.

At the time, however, no one was aware of this possible long-term political consequence, except perhaps for Stimson and then only in the abstract. Rather, immediate tactical considerations of war and peace prevailed, and all resources, even that of the atomic bomb, were applied to these short-term goals. In essence, the bomb was the ultimate symbol of power at the end of a long war in which the world had been caught up in a conflagration of power-mongering. In whatever sector the United States was seeking power, therefore, the bomb was seen as offering that power, whether militarily against the Japanese or politically against the Russians half a world away from where the bomb was actually dropped. It was because the bomb was seen in this light rather than within the context of future possibilities of it engendering Armageddon that, almost to a person, those involved in the decision-making were unanimous that the bomb be dropped. I believe Churchill is representative of both sides of the Atlantic in asserting that, 'There was unanimous, automatic, unquestioned agreement around our table; nor did I ever hear the slightest suggestion that we should not do otherwise.'[72]

C. SCIENTIFIC OPPOSITION TO THE USE OF THE ATOMIC BOMB

Although Margaret Gowing in her book, *Britain and Atomic Energy 1939-1945*, agrees with Churchill that 'whatever doubts there were among the British did not receive expression at the highest level', she does point to fairly intense scientific unrest and opposition to its use, particularly without warning.[73] Sir Henry Dale, President of the Royal Society, wrote to Churchill in 1944 asking him to see Niels Bohr because it was in Churchill's and Roosevelt's power 'even in the next six months to take decisions which (would) determine the future course of human history'.[74] Dale was concerned that the Allied leadership was approaching the question of the atomic bomb too uncritically, seeing it only in its utilitarian function to implement tactical gain. He felt that the sensitivity of someone such as Bohr, who was arguing for deeper thought about the long term consequences, could possibly communicate to Churchill the seriousness with which many scientists viewed the bomb. In essence, Bohr's position was that of a voluntary abstention from dropping the atomic bomb in the interests of future international control. When Bohr was finally given time to see Churchill, however, their meeting ended so disastrously that Churchill was to later write that, 'It seems to me Bohr ought to be confined, or at any rate made to see that he is very near the edge of mortal crimes.'[75] Whatever misgivings Sir John Anderson, Home Secretary, then Lord President, then Chancellor of the Exchequer, had, he never expressed them to Churchill, and when his Consultative Council was told in April 1945 that an atomic bomb was virtually ready for use, no voice was raised in protest.[76]

In America, a similar situation occurred, particularly among the scientists at the Metallurgical Laboratory at the University of Chicago where most of the Manhattan Project research was being done. Prominent among those apprehensive about what would happen to the world after their creation was given over to the politicians and the military was Leo Szilard. In 1939, as will be recalled, he and Einstein had encouraged Roosevelt to begin atomic research, for fear that Hitler might develop it first. In 1944, he had believed that unless the bomb was actually dropped, people would not understand it; by April 1945, however, once he began to realize the enormity of what the bomb could do, he changed his mind and sought to present his new beliefs to Truman. Instead, he was sent to see Byrnes with consequences equally as disastrous as that of the Churchill-Bohr meeting – Byrnes thought Szilard could be a spy.[77]

Nevertheless, Szilard persisted in airing his views, mostly among the Chicago scientists with whom he worked. The unrest was such that during the 31 May – 1 June meeting of the Interim Committee, Arthur Compton, Director of the Metallurgical Laboratory and member of the Scientific Panel, insisted that the growing dissidence at the Laboratory be addressed. The response of the Interim Committee was to set up several committees to discuss ideas on several topics, all of which would be incorporated into a single report. (This final report was submitted on 17 July.)

The most notable committee was the Committee on Social and Political Implications, whose chairperson was Professor James O. Franck. The most catalytic member, however, was Szilard, who tended to dominate the committee's final perspective. The committee stressed that the atomic bomb was of a different order from any other weapon in existence and that therefore 'the manner in which this new weapon is introduced to the world will determine in large part the future course of events'.[78] Within this perspective, the Committee was critical of their own work and urged the importance of *not* dropping the atomic bomb on Japan, particularly without a previous warning and demonstration. With the advent of nuclear weaponry, the report stated, safety among nations could only be sustained through international control. To drop the bomb on Japan, suddenly and unannounced, would greatly lessen future co-operation because of '. . . the ensuing loss of confidence and by a wave of horror and repulsion sweeping the rest of the world and perhaps even dividing public opinion at home'. The alternative suggested by the Committee (which Szilard again attempted to present to Truman) was to demonstrate the bomb in a place and in such a manner that would impress on all the world, particularly the Japanese, the horrendous devastation of its impact while not causing loss of life. This neutral demonstration, provided it was accompanied by a concomitant willingness on the part of the US to be the first nation to sacrifice some of its national sovereignty in order to do whatever was necessary to make atomic war impossible, would set in motion genuine feelings of trust among the nations at the close of a world war unprecedented in its hate and savagery. The report concluded by stating that only after such a neutral demonstration would an atomic bombing of Japanese cities be justified.

Known as the Franck memo, it was submitted to the Scientific Panel in the middle of June. The Scientific Panel remained unpersuaded, however, and stated in its report to the Interim Committee that:

> The opinions of our scientific colleagues on the initial use of these weapons are not unanimous. They range from the proposal of a purely technical demonstration to that of military application best designed to induce surrender. Those who advocate a purely technical demonstration would wish to outlaw the use of atomic weapons, and have feared that, if

> we use the weapons now, our position in future negotiations will be prejudiced. Others emphasize the opportunity of saving American lives by immediate military use, and believe that such use will improve the international prospects, in that they are more concerned with the prevention of war than with the elimination of this special weapon.
>
> We find ourselves closer to these latter views; we can propose no technical demonstration likely to bring an end to the war; we can see no acceptable alternative to direct military use.[79]

The Scientific Panel and the Interim Committee rejected the plea for a previous test, says Smith, because 'what mattered was that it would vitiate the shock effect on which the British and American leaders counted to undermine Japanese will to resist'.[80]

Still undaunted, Szilard wrote a petition in effect restating the Franck memo. He received sixty-nine signatures among his scientific colleagues and attempted to send it to the President himself, asking him to

> exercise your power as Commander-in-Chief to rule that the United States shall not resort to the use of atomic bombs in this war unless the terms which will be imposed on Japan have been made public in detail and Japan, knowing these terms, has refused to surrender. . . .[81]

On 17 July, Szilard turned the petition over to Compton for transmission 'through channels' to Washington and the President. There is no evidence that it ever reached Truman, who was by this time in Potsdam 'with a new feeling of confidence'.

What emerges at the onset of the development of atomic weaponry, then, is a bifurcation in analysis of the new device. Both political and military as well as scientific communities sought it while there remained the possibility that either the Germans or the Japanese could develop the bomb.[82] All recognized that whatever side gained mastery of the atomic weapon first would use it, and the thought of Hitler using it was unthinkable.

While the political and military communities continued to press for its development and use, viewing the atomic bomb as the perfect weapon for both military and political considerations, the scientists actually working on the construction of the bomb began to see that the weapon they were creating was not merely a new one in a series of others but a weapon of an entirely new order. They could only guess at its immediate impact and guess also at its long-term radiological consequences, but were impressed enough by its diabolic powers, even from the little they did know, to begin to pressure the governments of Britain and the US to consider a plan where it could be outlawed by international agreement without ever being used in battle. This feeling among scientists solidified increasingly after it became apparent that neither Germany nor Japan were going to develop their own atomic bombs.

The general political situation was grim, however, and the idealism of the scientists did not fit into the Allied impatience to end the strain of war as quickly as possible. Just as Niels Bohr was simply unable to focus Churchill's mind, preoccupied as it was at the time with the invasion of Europe, on the long-term implications of dropping the bomb, so the scientists in America met a similar fate, particularly with Truman. Things might have turned out differently had there been the opportunity for full-scale political debate through the normal democratic channels of public opinion and parliamentary representation, but the entire project was classifed top secret so that while the impact of the bomb was to be *global*, the number of persons entering into making the fateful decision whether or not to drop it was no more than *several dozen*. And the number of persons who even *knew* about it prior to 6 August was no more than several hundred. The development of the atomic bomb was one of the best kept secrets of the war.

D. TRINITY

Perhaps the greatest indication of the ineffectiveness of the opposition to dropping the atomic bombs on Japan can be seen in the fact that at no point was research and development of the bomb slowed down or questioned. Even the scientists involved in the campaign of Szilard continued to work on the bomb while they asked that it should not be dropped. Under the command of Major General Groves and the scientific direction of Oppenheimer, research and production continued feverishly, and on 16 July, while Truman and Stimson were away at Potsdam, the testing of the first atomic bomb in history took place. Codenamed Trinity, the test involved the detonation of a ball of plutonium the size of a grapefruit set on top of a hundred foot tower in a remote and forbidding area of the New Mexico desert called Jornada del Muerte – Journey of Death.

At 5:29:45 a.m. the device exploded with a force equivalent to 20,000 tons of TNT, digging a crater 1,200 feet in diameter 'from which all vegetation vanished, and vapourizing a four-inch iron pipe 1,500 feet away. At Ground Zero the temperature at the moment of detonation was one hundred million degrees Fahrenheit, three times hotter than the interior of the sun and 10,000 times hotter than the surface of the sun. A big boom came about 100 seconds after the great flash – 'the first cry of a new-born world', as William Lawrence remembers it.[83]

Oppenheimer, watching from a bunker eleven miles away, remembers seeing 'an unbelievable light' and recalls that

> we waited until the blast had passed, walked out of the shelter and then it was very solemn. We knew the world would not be the same. A few people laughed, a few people cried. Most people were silent. I remembered the line from the Hindu scripture, the Bagavad Gita: Vishnu is trying to persuade the Prince that he should do his duty and to impress him takes on his multi-armed form and says, 'Now I am become death, destroyer of worlds.' I suppose we all thought that one way or another.[84]

One in particular who did was George Kistiakowsky, who remembers the experience as 'the nearest thing to doomsday that one could possibly imagine. I am sure that at the end of the world – in the last millisecond of the earth's existence – the last man will see what we have just seen.'[85]

The reaction of Groves was more concrete: 'The war's over. One or two of these things and Japan will be finished.'[86]

The following message was immediately relayed to Truman in Potsdam: 'Operated on this morning. Diagnosis not yet complete but results seem satisfactory and already exceed expectations.'[87] Truman read the message, realizing, he recalled later, that 'the most secret and the most daring enterprise of the war had succeeded. We are now in possession of a weapon that would not only revolutionize war but could alter the course of history and civilization.'[88] Stimson met Churchill the next morning, and Churchill responded with his characteristic eloquence: 'Stimson what was gunpowder? Trivial. What was electricity? Meaningless. This atomic bomb is the Second Coming in wrath.'[89] Stalin was told nothing, Truman only 'casually' mentioning to him on 24 July, at the close of the conference, that the US had 'a new weapon of special destructive force'.[90]

E. HIROSHIMA

On 24 July 1945, President Truman ordered the atomic bombing of Japan: 'The 509th Composite Group, 20th Air Force, will deliver its first special bomb as soon as weather will permit visual bombing after about 3 August 1945, on one of the targets: Hiroshima, Kokura, Niigata and Nagasaki. . . .'[91]

On 26 July the Allies issued a renewed call, the Potsdam Declaration, for Japan to surrender unconditionally. On 28 July, while still attemping to negotiate through the Russians, the Japanese rejected the Declaration, Prime Minister Suzuki saying that it was 'unworthy of public notice'.[92]

Nearly a year previously, on 1 September 1944, the 393rd Heavy Bombardment Squadron of the US air force, based in Nebraska, had been assigned to Lieutenant Colonel Paul Tibbets and directed to begin practice runs to prepare for dropping the bomb. By the time of the final days between Potsdam and 6 August, Tibbets and his crews were on stand-by on the Trinian islands, southwest of Japan, from where the bombs were to be delivered.

Although Groves had wanted to bomb Kyoto first, Stimson overruled him, stating that while it was a military target of considerable importance, it had been the ancient capital of Japan and was therefore a shrine of Japanese culture and art. Instead, Stimson approved four other target sites, including Hiroshima and Nagasaki, both 'active working parts of the Japanese war effort. Hiroshima was not only a major assembly and military storage point but was the headquarters of the Japanese army defending southern Japan. Nagasaki was a major seaport with several large industrial plants'.[93]

After several days delay because of bad weather, the bombing mission finally took off from Trinian towards Japan at 2:47 a.m. on 6 August. The bomb was aboard the Enola Gay, commanded by Tibbets, and was accompanied by two observation planes carrying cameras, scientific instruments and trained observers. At 3:00 assembly of the bomb began; by 3:20 this was completed, making it a 'final bomb'.[94] Known by its makers as Lean Boy or Thin Boy, the atomic bomb contained U^{235} as its fissionable material. The U^{235} was enclosed at the opposite ends of a gun barrel fifty-two inches long. To be ignited, both sections were to be brought together in a critical mass and exploded by a proximity fuse when the gun mechanism shot the smaller of the two uranium parts into the other. When the time came for detonation, therefore, the detonator would ignite a charge of high quality gunpowder which would propel the smaller uranium slug down the barrel at 900 feet per second towards the second uranium ring. As this 'bullet' travelled down the barrel, a small device made of Polonium 84 would begin emitting neutrons, thus initiating the chain reaction.[95]

Tibbets had been ordered to scrap the mission if Hiroshima was sealed in clouds. At 7:24 a.m., however, his weather plane radioed: 'Cloud cover less than 3/10ths at all altitudes. Advice: bomb primary.'[96] This report, according to the official US air force history, 'sealed the city's doom'.[97]

To appreciate fully what happened next two things should be borne in mind. The first is the magnitude of what the Enola Gay was carrying. Most of the destruction that occurred during World War II was caused by bombs having the equivalent of about one ton of TNT. Towards the end of the war, new bombs were developed containing ten tons of TNT; they were termed 'blockbusters' because of their amazing capacity to devastate an entire city block. The very biggest of these blockbusters contained almost twenty tons of TNT, the largest one being the 'Grand Slam', made in Britain.[98]

The bomb that was dropped on Hiroshima exploded with a force equivalent to 20,000 tons of TNT or *one thousand* times the force of the largest blockbuster. Frank Clune puts the comparison another way by pointing to the firebombing of Dresden in 1944, perhaps the worst bombing event of the European theatre of the war. Over 60,000 people were killed in an attack composed of over fifteen hours of continuous saturation bombing by Allied planes. The Hiroshima bomb pulverised an entire city in one single blast.[99]

The second thing to be kept in mind is that almost inversely proportional to the *magnitude* of the bomb was the city's *unpreparedness* for it. It was wartime, of course, and people had expected a conventional bombing raid to occur at some point, but though air raids had been regularly sounded, only an occasional stray bomb had ever been dropped on Hiroshima. It is true that on 27 July American planes dropped leaflets warning Hiroshima residents that their city was going to be demolished if the Japanese did not surrender immediately and unconditionally, but research indicates that very few people ever saw these leaflets and those who did discounted them as enemy propaganda.[100] In any case, the leaflets made no mention of an atomic bomb or any other kind of special weapon. As mentioned previously, this omission was a conscious one on the part of the American decision makers, as it was felt that any previous warning of a specifically nuclear attack 'would hinder rather than serve our wish to end the war at once'.[101]

There was a more immediate factor, that contributed to the element of total surpise. The night previous to the bombing had witnessed two separate air-raid alerts during which no bombs had fallen. At 7:10 a.m. on 6 August a third alert sounded, following the spotting of Commander Tibbets' weather planes over southern Japan. Because of these three false alarms no additional alert was sounded when the actual bombing planes appeared over the city shortly after 8:00 a.m. The only warning was that of a radio announcement noting that the planes appeared to be on a reconnaissance mission.

Since people began their workdays early during the wartime summer, many were already at work or en route by 8:00 a.m. Housewives were completing after-breakfast chores with the charcoal still burning in their hibachis, the general atmosphere being that of early morning complacency. Hiroshima was quite simply as in the days of Noah before the flood: 'They were eating and drinking, marrying and giving in marriage, until the day when Noah entered the ark, and they did not know until the flood came and swept them all away. . . .' (Matt. 24.38-39a).

As one survivor later described it:

> The sky was serene, the air was flooded with glittering morning light. My steps were slow along the dry, dusty road. I was in a state of absent-mindedness. The sirens and also the radio had just given the all-clear signal. I had reached the foot of the bridge, where I halted, and was turning my eyes toward the water. . . .[102]

Precisely at 8:15:17 a.m. the bomb-bay doors of the Enola Gay snapped open. The plane lurched upwards nearly ten feet, suddenly 9,000 pounds lighter. Inside the bomb a timer tripped the first switch in the firing circuit, letting the electricity travel a measured distance towards the detonator. At 5,000 feet above the ground a barometric switch was triggered. At 1,890 feet the detonator was activated, the bomb exploding at exactly 8:16, forty-three seconds after falling from the Enola Gay.

In the first milli-second after detonation, a pinprick of purplish-red light expanded into a glowing fireball a half mile in diameter; the temperature at its core was 50,000,000 degrees fahrenheit. At Ground Zero, some two thousand feet directly underneath the explosion, temperatures reached several thousand degrees, melting the surface of granite 1,000 yards away. Nine hundred yards from the epicentre, several thousand soldiers, including one American prisoner of war, were doing their morning exercises in the courtyard of Hiroshima Castle: they were instantly incenerated, their charred bodies burnt into the ground.

Watching from below, one child remembered later the entire sequence leading up to this first milli-second: 'I was watching the aeroplane the whole time. . . . Suddenly a thing like a white parachute came falling. Five or six seconds later everything turned yellow in one instant. It felt the way it does when you get the sunlight straight into your eye. . . .'[103]

After this first blast of light, the fireball suddenly exploded into a mass of swirling flames and purple clouds out of which came a huge column of white smoke which rose to a level of 10,000 feet. At 10,000 feet, the column flattened outwards to form a mushroom-shaped cloud, then climbed upwards until it reached a height of 45,000-50,000 feet.

Aboard the Enola Gay, observing the scene at an altitude of 29,000 feet eleven miles away, the scene was quite spectacular as the tailgunner, Sergeant George Caron, recalls:

> A column of smoke rising fast. It has a fiery red core. A bubbling mass purple-grey in colour, with that red core. . . . It's all turbulent. Fires are springing up everywhere, like flames shooting out of a huge bed of coals. . . . Here it comes, the mushroom shape . . . like a mass of bubbling molasses. . . . It's very black, but there is a purplish tint to the cloud. The base of the mushroom looks like a heavy undercast that is shot through with flames. The city must be below that. . . .

It was, Caron remembered later, like 'a peep into hell'.[104]

One person in the 'hell' Caron describes remembers things much more starkly, recalling only 'a sudden flash, a blast, and then a cataclysmic earthquake – all the representatives of disaster and death, each following the other'.[105] Much more eloquent is the recollection of another survivor who remembers it as being awesome and replete with diabolic beauty:

> A blinding . . . flash cut sharply across the sky . . . I threw myself onto the ground . . . in a reflex movement. At the same moment as the flash, the skin over my body felt a burning heat . . . (then there was) a blank in time . . . dead silence . . . probably a few seconds . . . and then a . . . huge boom . . . like the rumbling of distant thunder. At the same time a violent rush of air pressed down my entire body. . . . Again there were some moments of blankness . . . then a complicated series of shattering noises . . . I raised my head, facing the centre of Hiroshima to the west . . . (there I saw) an enormous mass of clouds . . . (which) spread and climbed rapidly . . . into the sky. Then its summit broke open and hung over horizontally. It took the shape of . . . a monstrous mushroom with the lower part as its stem – it would be more accurate to call it the tail of a tornado. Beneath it more and more boiling clouds erupted and unfolded sideways . . . the shape . . . the colour . . . the light . . . were continuously shifting and changing.[106]

The aiming point of the bomb was a central area of Hiroshima adjacent to an Army Headquarters with 60% of the population within 1.2 miles; the drop was accurate to within 200 metres of the aiming point. The explosion created an area of total destruction extending two miles in all directions in a flat city of homes and buildings made largely of wood. Within three miles of the epicentre 62,000 of 90,000 buildings were destroyed. All utilities and transportation services were demolished, and twenty-six of thirty-three fire stations were destroyed, leaving only sixteen pieces of fire-fighting equipment to deal with a whole city on fire. Forty-two of forty-five of the city's hospitals were devastated, killing 270 of 298 doctors and 1,645 of 1,780 nurses.[107]

The number of deaths, both immediately and after a period of time, will never be known. They are variously estimated from 63,000[108] to 240,000.[109] The official US estimate is 78,000, although the city of Hiroshima estimates 200,000.[110] The reason for this enormous disparity is due to such things as differing techniques of calculation, varying estimates of the actual number of people in the city at the time, the manner in which military fatalities are included, and after which census count the approximation was made. It is safe to say that of a city population variously estimated to be from 270,000 to 400,000, between 25% and 50% were killed by one solitary fifteen kiloton atomic bomb.

It is also safe to say that the entire city became immediately involved in the atomic disaster. Within 1,500 feet of the epicentre, research has concluded there were 3,483 people; 88% died either instantly or that same day, only fifty-three survived beyond a few years. If a person survived within 1,000 metres, more than 90% of the people around him or her were fatalities; if a person was unshielded at 2,000 metres, 80% were killed. Although mortality was lower if one was indoors, one had to be at least 2,200 metres (1.3 miles) from the epicentre to have even a 50% chance of surviving immediate death. The heaviest casualties were among the children.[111]

So complete was the destruction that a history professor recalls that

> I climbed Hijiyama Hill and looked down. I saw that Hiroshima had disappeared. . . . I was shocked by the sight. . . . What I felt then and feel now I just can't explain with words. Of course I saw many dreadful scenes after that – but that experience, looking down and finding nothing left of Hiroshima – was so shocking that I simply can't express what I felt. I could see Koi (a suburb at the opposite end of the city) and a few buildings standing. . . . But Hiroshima didn't exist – that was mainly what I saw – Hiroshima just didn't exist.[112]

'Such a weapon', he was to recall later, 'has the power to make everything into nothing.'[113]

F. NAGASAKI AND SURRENDER

The press statement issued from Washington, DC on the afternoon of 6 August announced that a bomb had been dropped on Hiroshima with a yield of 20,000 tons of TNT. In simple and concise language it stated:

> It is an atomic bomb. It is a harnessing of the basic powers of the universe. The force from which the sun draws its power has been loosed against those who brought war to the Far East.

The statement concluded by warning:

> It was to spare the Japanese people from utter destruction that the ultimatum of July 26 was issued at Potsdam. Their leaders promptly rejected that ultimatum. If they do not now accept our terms, they may expect a rain of ruin from the air, the like of which has never been seen on this earth.[114]

When the Japanese Emperor heard the news from Hiroshima he said:

> Under these circumstances, we must bow to the inevitable. No matter what the effect on my safety, we must put an end to this war as speedily as possible so that this tragedy will not be repeated.[115]

Although the Japanese Emperor wanted to surrender at this point, his generals remained in favour of the war effort, still arguing that a decisive battle could be won that would ensure a negotiated settlement.[116] They further urged that the effects of the atomic bombing be minimized. Without a large enough constituency supporting him, the Emperor relented, and on the morning of 7 August Tokyo radios reported that 'A small number of B-29s penetrated into Hiroshima city a little after 8:00 a.m. yesterday morning and dropped a small number of bombs. As a result a considerable number of homes were reduced to ashes and fire broke out in various parts of the city.'[117]

On 8 August, while the US intensified its call for immediate unconditional surrender, 'otherwise we shall resolutely employ this bomb and all our superior weapons to promptly and forcefully end the war',[118] the Japanese military continued to insist that the war be carried on and the devastation of Hiroshima be minimized. That afternoon, with no Japanese surrender forthcoming, Top Secret Field Order 17 was issued, naming Kokura as the primary target, Nagasaki the secondary target, for the next atomic bombing.

At 10:58 a.m. on 9 August, as the Supreme War Council continued to argue in Tokyo, Major Charles Sweeney, commander of the second atomic bombing mission, began his final run over Nagasaki, having been forced to relinquish Kokura because of weather conditions. Although the bomb used against Hiroshima was of Uranium 235, the Nagasaki bomb was of Plutonium 239. The type of damage produced by both were similar, however, both explosions resulting in massive damage and various combinations of mechanical, thermal and radiation death and/or injury.

Like the Hiroshima bomb, the moment of detonation was an explosion of light exceeding 300,000 degrees centigrade. Frank Chinnock describes the impact of this first milli-second of heat:

> For some 1,000 yards, or three fifths of a mile, in all directions from the epicentre . . . it was as if a malevolent god had suddenly focused a gigantic blowtorch on a small section of our planet. Within that perimeter, nearly all unprotected living organisms – birds, insects, horses, cats, chickens – perished instantly. Flowers, trees, grass, plants, all shrivelled and died. Wood burst into flames. Metal beams and galvanised iron roofs began to bubble, and the soft gooey masses twisted into grotesque shapes. Stones were pulverized, and for a second every last bit of air was burned away. The people exposed within that doomed section neither knew nor felt anything, and their blackened, unrecognizable forms dropped silently where they stood.[119]

After this explosion of heat and light came the explosion of sound, travelling at a speed of 9,000 miles per hour. It destroyed every building within 800 metres, and of the 55,000 buildings throughout the city, 20,000 were demolished. More would have been damaged were it not for the hilly topography of Nagasaki which kept much of the destruction confined to Urakami Valley in the northern sectors of the city.

Immediately after the explosion of sound came the third and final killer: radiation, probably the most frightening of all because it acted silently and invisibly and at the time was barely even known about, much less understood. At the moment of the explosion, various radiation rays were emitted: beta, gamma, alpha and X-rays and neutrons. The X-rays, gamma rays and neutrons were most injurious. In addition, fission products such as strontium 90, which attacks the white blood cells, causing leukemia, and cesium 137, which is carcinogenic to muscle tissue and the ova of females, were scattered everywhere. These were later named by the survivors of both Hiroshima and Nagasaki the 'Ashes of Death', because they constituted the 'invisible contamination' which struck the people without warning and with no hope of cure.

The effects of the radiation exposure were felt both immediately and long afterwards. If a human body receives 400-500 roentgens of radiation it has only a 50% chance of survival: if it receives 700-800 roentgens, the fatality rate is virtually 100%. Those within 1,000 metres of Ground Zero received well above this dosage, although, as with the heat, the percentage of those who felt the effects of the rays decreased as the distance increased from the epicentre. A useful comparison of the amount of radiation expelled and received, however, can be seen when it is recalled that at the Three Mile Island nuclear reactor near Harrisburg, Pa., pregnant women and pre-school children were evacuated after a release of merely 0.01 roentgen units. This is to say, that the radiation released from both the Hiroshima and Nagasaki bombs was so massive that even without the heat blast or the explosion of sound the number of deaths within 1,000 metres of the epicentre would have remained essentially the same. The primary difference would have been in the amount of time it took the victims to die. Those killed instantly would have lived instead for twenty-four to forty-eight hours, until the radiation destroyed their white blood cells and bone marrow, causing death from radiation sickness. A Dr Raisuke Shirabe of the Nagasaki Medical College describes the effects of the radiation in a way that gives an account of the symptoms while illuminating the degree of ignorance concerning the sickness among the Japanese medical communities:

> Two days have passed since the big bomb was dropped on us, and I still can find no explanation as to why people continue to die. Many of them have no visible injuries, yet most exhibit the same symptoms: bleeding from the gums, loss of appetite, fever, apathy, the beginning of loss of hair, bloody diarrhoea – then death. What is this strange, invisible killer that remains in our midst?[120]

Perhaps the most insidious characteristic of radiation is that the younger one is the more vulnerable one is to its effects, particularly children *in utero*. All recorded cases of pregnant women within 3,000 metres of Ground Zero resulted in miscarriages or premature births at which time the infants died. Pregnant women beyond 3,000 metres suffered a 66% miscarriage and stillbirth rate.

The overall estimation of casualties from the Nagasaki bombing, like at Hiroshima, varies greatly. The US official estimate is 40,000 killed; the Nagasaki estimate is 74,800 killed, 75,000 wounded. In all cases, what is certain is that Nagasaki, like Hiroshima, was immediately inundated by death and destruction on an unprecedented scale.

Even after this second cataclysm, however, the Japanese Supreme War Council wished to fight on. Finally, at 2:00 a.m. on 10 August, after a grim and protracted debate in which the War Minister, joined by the Chiefs of Army and Navy Staff, argued that Japan could get better terms than the Potsdam Ultimatum if she hung on, the Prime Minister, Suzuki, stepped forward and in an act unprecedented in modern Japanese history asked the Emperor to decide. This issuance of an absolute command by the Emperor was known among the Japanese as the Voice of the Sacred Crane. When it was heard, it called for immediate unconditional surrender.[121]

G. HIBAKUSHA: 'SURVIVOR AS EVERYMAN'

Although tone is so integral a part of the self that one does not so much decide upon it as feel impelled towards it as the matrix of one's moral and intellectual formulation, Robert Lifton asserts that 'there is no tone, no framework, adequate to the nuclear weapons experience'.[122]

Lifton makes this claim, based on two years of research, conducted in Tokyo and Kyoto from 1960-62, on the relationships of the individual character and historical change in Japanese youth. His studies indicated that the most fundamental characteristic among the youth is a sense of being cut off from their past, a cut which was in some sense bound up with the Japanese encounter with the atomic bombs dropped upon them at the close of World War II. Because of the impact of the atomic bomb upon Japanese youth generally, Lifton began a systematic study of atomic bomb survivors, producing the classic study *Death in Life: The Survivors of Hiroshima*. This work has been the major source for the present section, although numerous other studies have been written on the subject.[123]

Hibakusha is a coined word which literally means 'explosion-affected person'; it is used to refer to all those who survived either Hiroshima or Nagasaki atomic bombings. According to official definition, the category of hibakusha includes: (1) those within the city limits at the time of the attack, an area extending from the bomb's epicentre to 4,000, sometimes 5,000 metres; (2) those who were not in the city at the time but who returned to within 2,000 metres of the epicentre within fourteen days; (3) those who engaged in some form of aid to, or disposal of, bomb victims; and (4) those who were *in utero* and whose mothers fit into any of the first three categories.

Just as Lifton observed that all Japanese, whether officially hibakusha or not, were impacted by the atomic bomb experience, particularly the youth, so all of us are in some degree affected as well; that is to say, that the psychological occurrences in Hiroshima and Nagasaki have important bearing on all of human existence. What happened then can happen again to all of us. As Lifton puts it: 'We are all survivors of Hiroshima and, in our imaginations, of future nuclear holocaust.'[124] This is a link which is not metaphorical but real, both because of the omnipresent spectre of this holocaust in fact happening and because the specific psychological components of the Hiroshima hibakusha can be explored in relationship to the general psychology of the survivor, a category in which the rest of humanity since 6 August can be placed.

That we are all survivors can be seen from the definition of survivor Lifton offers, namely 'one who has come into contact with death in some bodily or *psychic* fashion and has himself remained alive'.[125] I have emphasized psychic because that is the category in which all non-officially hibakusha fall. It is the category which must be thoroughly explored in order to comprehend the magnitude of the Hiroshima experience upon human history and psychic life: psycho-history, if you will.

Perhaps this psychic dimension has been most eloquently described by one particular hibakusha, Pedro Arrupe. Now Father General of the Society of Jesus, he was near Hiroshima when the bomb was dropped. He fits into the third category of hibakusha since he was a doctor and spent weeks in the city caring for the victims. He details his impression as a witness to that event by what the initial blast did to his wall clock: it stopped it. The Hiroshima event, he says, did precisely that to the entire historical process: it stopped it. 'For me,' he writes in his book, *A Planet to Heal*, 'that silent and motionless clock has become a symbol. The explosion of the first atomic bomb has become a para-historical phenomenon. It is not a memory, it is a perpetual experience, outside history, which does not pass with the ticking of the clock. The pendulum stopped and Hiroshima has remained engraved on my mind. It has no relation to time. It belongs to motionless eternity.'[126]

By exploding the normal cyclic patterns through which the human family is accustomed to living and dying, the atom bomb placed us all in an altered state of space/time, revolutionizing crime and punishment and demanding a corollary transformation in our conception of world justice to balance our new-found powers to meet out destruction. Quite literally Arrupe recalls his consciousness being seared by the atomic bomb, cutting him off from his past and placing him within the perpetual context of death immersion. His experience is similar to the hibakusha Lifton studied, people who were all caught up in 'the sense of a sudden and absolute shift from normal experience to an overwhelming encounter with death'.[127]

In an examination of the hibakusha, what will emerge are the themes of death imprintation, death guilt, psychic numbing, counterfeit nurturance and contagion, and formulation. These are all universal psychological tendencies, ones which all of us to some degree participate in as we try to reconcile the omnipresent possibility of nuclear holocaust with our personal lives and as we attempt to put time back in motion and space back in order after the Hiroshima bomb exploded all our known categories. We are all survivors of Hiroshima; we are all potential victims of its re-enactment.

In order to understand the full impact of Hiroshima, detached intellectual appreciation is not enough. The demand is that the 'observer' realize that s/he is *part of that experience*, that the same psychological reactions and formulations occurring then and there are playing a role in each of us here and now. Our connection, therefore, must be seen as *organic*; organic because Hiroshima has deeply impacted our very psyches. With Arrupe we must realize that Hiroshima has remained 'engraved' on the human mind since 6 August 1945. In the examination of the hibakusha in this next section, therefore, it is fundamental to realize that their experience is paradigmatic of all of us: 'the survivor becomes Everyman'.[128]

1. *Death Immersion*

The experience of one woman hibakusha is expressive of most in the immediate moments after the detonation of the Hiroshima bomb:

> Immediately following the flash there was a tremendous explosion. A mighty air wave seemed to lift me upward. In

> the next moment, I was being crushed against the ground by ponderous pressure. The force of the explosion raised a cloud of swirling black dust mixed with wood fragments and slivers of broken glass. All of us were injured or cut.[129]

The intensity of the dislocation and shock can be seen in the widespread belief many held that what they were experiencing was not another bombing or wartime attack but the destruction of the entire world. A physicist, temporarily blinded and covered by falling debris, recalled: 'My body seemed all black, everything seemed dark, dark all over . . . then I thought, "The world is ending." '[130] Similarly a Protestant minister: 'The feeling I had was that everyone was dead. The whole city was destroyed . . . I thought all of my family must be dead – it doesn't matter if I die. . . . I thought this was the end of Hiroshima – of Japan – of humankind. . . . This was God's judgment on man'.[131] Another religious domestic worker stated that his immediate reflection was that 'there is no God, no Buddha'.[132] This seemed plausible to many as devastation was not localized but stretched as far as any of them could see. Indeed, as Dr Hachiya described those few moments after the blast, 'Hiroshima was no longer a city but a burnt-over prairie'.[133]

Beyond the death imagery *per se* that seemed to envelope the survivors, much of it epitomized by olfactory images – a 'smell of death', there followed a widespread sense that life and death had been blown out of synchronicity with one another, that they were no longer properly distinguishable – a realization that gave an atmosphere of macabre unreality. Part of this unreality was what many hibakusha termed a 'deathly silence' which followed the attack. Rather than hysterical panic and confused pandemonium, most remember a numbing stillness, of being thrust from normal living in after-breakfast Hiroshima into a state of suspended animation or slow motion: a 'complete silence', as Dr Hachiya recalls.[134]

The only sounds were the moanings of those not yet dead; the rest slowly moved from the destruction towards the rivers where they thought their families might be, or towards other accumulations of the living. In many cases, people recall moving along with a gathering human mass of barely audible slowness and with no apparent destination. As one hibakusha remembers it, it 'was like taking a stroll through the lowest reaches of hell'.[135] Dr Hachiya describes the scene in more detail:

> Those who were able walked silently toward the suburbs in the distant hills, their spirits broken, their initiative gone. When asked whence they had come, they pointed to the city and said, 'that way'; and when asked where they were going, pointed away from the city and said, 'This way'. They were so broken and confused that they moved and behaved like automatons.
>
> Their reactions had astonished outsiders who reported with amazement the spectacle of long files of people holding stolidly to a narrow rough path when close by was a smooth, easy road going in the same direction. The outsiders could not grasp the fact that they were witnessing the exodus of a people who walked in the realm of dreams. . . . A spiritless people had forsaken a destroyed city. . . .[136]

One of these 'automatons' walking in the 'realm of dreams' and, as Dr Hachiya also recalls, smelling 'like burning hair', was a watch repairman. He remembers that

> all the people were going in that direction and so I suppose I was taken into this movement and went with them. . . . I couldn't make any clear decision in a specific way . . . so I followed the other people. . . . I lost myself and was carried away. . . .[137]

A grocer, himself badly burned, conveys this profound death-life disruption and sense of death-in-life reality immediately after the bombing by recalling the physical appearance of these wandering people:

> The appearance of people was . . . well, they all had skin blackened by burns . . . they had no hair because their hair was burned, and at a glance you couldn't tell whether you were looking at them from in front or in back. . . . They held their arms bent (forward) . . . and their skin – not only on their hands, but on their faces and bodies too – hung down. . . . If there had been only one or two such people . . . perhaps I would not have had such a strong impression. But whenever I walked I met these people. . . . Many of them died along the road – I can still picture them in my mind – like walking ghosts. . . . They didn't look like people in this world . . . they had a special way of walking – very slowly. . . . I myself was one of them.[138]

Perhaps the most common injury was that of burns, and the grotesqueness of this is often what survivors remember most clearly. For example, one hibakusha recalls that 'outside I saw people dragging what at first looked like white cloth but what I later saw was skin that had peeled from their bodies'.[139]

Most in fact did not survive, for as a Dr Tabushi remembers, 'all through the night, they went past our house, but this morning they had stopped. I found them lying on both sides of the road so thick that it was impossible to pass without stepping on them'.[140]

The phrase used repeatedly by many of those who did survive is 'muga-muchu', literally meaning 'without self, without centre'. As the hibakusha recall it, *the atomic attack quite simply exploded the boundaries of the self*. What the hibakusha found themselves thrust into, therefore, both in terms of the immediate impact of the bomb as well as the longer-term physical and psychological consequences, was, to quote Lifton, *'a vast breakdown of faith in the larger human matrix supporting each individual life, and therefore a loss of faith (of trust) in the structure of human existence'*.[141]

This 'vast breakdown' was due primarily to what Lifton terms the 'indelible imprint of death immersion'.[142] This immersion involves a complex of many factors: the fear of annihilation of both self and individual identity after having virtually experienced that annihilation; the destruction of the ecological environment; the devastation of the field of one's normal existence and hence of one's general sense of 'being-in-the-world';[143] and finally and most importantly, what Lifton calls the *'replacement of the natural order of living and dying with an unnatural order of death-dominated life'*.[144]

The key to understanding the Hiroshima experience, then, both in terms of its immediate devastation and in terms of its impact on the psyche of the survivors, is by way of the overwhelming imprint of death on both environment and people. It is an imprint that permanently 'engraved' in their minds the imagery of death. This imprint has been a lasting one in the life of the hibakusha for two reasons: (*a*) that of the long-term impact of radiation upon the *body*, affecting even future generations; and (*b*) the lasting impact of the Hiroshima event upon the *psyche* of the hibakusha.

(a) Physical Death Immersion: The 'Invisible Contamination' of Radiation

Soon after the bomb fell, survivors began to notice signs of an illness no one could properly understand or diagnose. It consisted of nausea; vomiting and a general absence of hunger; diarrhoea mixed with blood; fever and an overall sensation of bodily weakness; purple spots on various parts of the body from bleeding into the skin (purpura); ulceration and inflammation of the mouth cavity and throat (oropharyngeal lesions and gingivitis); bleeding in the mouth, gums, throat, urinary tract and rectum (hemorrhagic manifestations); loss of hair (epilation); and an extremely low white blood cell count (leukopenia).

It was not a sickness that manifested itself all at once; rather, it appeared gradually: gastrointestinal symptoms appearing first, then the hemorrhagic and bone-marrow effects hours or weeks later, depending on the dosage of radiation received.

While it seems obvious to us that this is radiation sickness, at the time, because an atomic bomb had never been dropped, radiation sickness was virtually unknown. Dr Hachiya, Director of the Hiroshima Communications Hospital, for example, was not certain of the true nature of the disease until he attended a lecture by two outside consultants on 3 September. At first he thought it was a poison gas or germ bomb, and it was not until 12 August that he knew that it had been an atomic bomb that had been dropped.[145] The difficulty in making what appears to us now as an obvious connection was due, according to one hibakusha, to the fact that 'ordinary bombs just destroy in a visible manner, but the A-bomb destroys people invisibly'.[146]

Because the bomb destroyed people 'invisibly', the impact upon the survivors was as devastating emotionally as physically, for the bomb held for them a special terror other diseases did not hold: it yielded, to quote Lifton, *'an image of a weapon which not only instantly kills and destroys on a colossal scale but also leaves behind in the bodies of those exposed to it deadly influences which may emerge at any time and strike down their victims'*.[147]

This image was made particularly vivid by the fact that two to four weeks after the bombing itself, people who had up till then seemed in perfect health succumbed to the radiation sickness and died. The eerie quality of radiation sickness is made particularly clear by the comments of an electrician:

Those sick people . . . didn't seem in pain. Only they

> couldn't move, and even as we watched them they seemed to become faint. . . . And even those who looked as though they would be spared were not spared. . . . People seemed to inhale something from the air which we could not see . . . the way they died was different . . . and strange.[148]

Even the proper naming of the disease did not lessen the terror the people felt, according to a Buddhist priest:

> We heard the new phrase, 'A-bomb disease'. The fear in us became strong, especially when we would see certain things with our eyes. A man looking perfectly well as he rode by on a bicycle one morning, suddenly vomiting blood, and then dying. . . . Soon we were all worried about our health, about our bodies – whether we would live or die. And we heard that if someone did get sick, there was no treatment that could help. We had nothing to rely on, there was nothing to hold us up. . . .[149]

People, terrorized by the spectre that they, too, could be victimized, uncertain as to source or contagion possibilities and mystified as to its sudden manifestations and apparent incurableness, began to yield to a strong unconscious wish to separate themselves from persons they saw to be afflicted. And yet they knew that they were not safe themselves, and so the mere act of separation from those already stricken gave rise to severe senses of guilt.

This sense of guilt-ridden mistrust and self-doubt was not to subside after days or weeks, for the effects of the radiation never completely subsided; indeed, the physical fears experienced in relationship to early radiation effects have turned into lifetime bodily concerns which are still very much with those hibakusha.

After the initial bout with radiation sickness, the hibakusha still living, if they survived, were confronted by still another delayed impact: abnormally high rates of leukemia. The increased rate of this disease was first noted in 1948, although it did not reach a peak until 1950-52. Greatest hit were those hibakusha closest to the epicentre, primarily those within 2,000 metres; for those within 1,000 metres, the incidence of leukemia has been between ten and fifty times the normal rate. Since 1952, the rate has gradually decreased but it is still running at a higher-than-normal rate and fears among the hibakusha remain strong.[150]

After increases in the leukemia rates came increases in cancer rates, particularly carcinoma of the thyroid, lung, stomach, ovary and uterine cervix.[151] These were not noticed until after the increases in leukemia because cancer has a longer latency period. There has also been higher-than-normal incidence of cataracts and other eye-related conditions, most appearing within two years of exposure. Other divergent conditions, without as much clear-cut scientific evidence as that of leukemia and cancer, but thought by many physicians and most hibakusha to result from the bomb, include: central nervous-system impairment, particularly in the mid-brain, sexual dysfunction, several types of anemia and other blood diseases, endocrine and skin disorders, premature aging, liver diseases and, perhaps most difficult of all to evaluate, a report by most hibakusha of feeling a general weakness and debilitation.

There is still another even more devastating consequence of the 'invisible contamination' of radiation exposure, however: the impact on the unborn, not only upon those *in utero* during the blast but upon those conceived years later.[152] For those *in utero*, the results were a dramatic increase in stillbirths and abortions. There was also among those under four months a high incidence of microcephalia both with and without mental retardation.[153] Reports from hibakusha indicate that other congenital defects and childhood abnormalities are occurring among children born to them, but unfortunately not enough systematic studies have been done to prove this scientifically. The popular attitude in Japan believes this to be the case, however, with or without the scientific studies, and hibakusha have encountered enormous discrimination, not only in occupational areas but in marital arrangements as well.[154]

'A-bomb disease', then, represents for the hibakusha an endlessly devastating phenomenon that plagues not only their own lives but those of their children. The time since the bombing has become an indefinite extension of the 'invisible contamination' so many died from immediately after the blast. Survivors see themselves as *continually susceptible* to a bodily destruction that *began* on 6 August 1945, but which has yet to end. To most hibakusha, 'A-bomb disease' has a ring of perpetual fatality which they cannot elude nor spare their children from.

As one might expect from this, the omnipresent spectre of being stricken by 'A-bomb disease' has given rise to 'A-bomb neurosis'. This Lifton defines as a 'precarious inner balance between the need for symptoms and the anxious association of these symptoms with death and dying'.[155] Because the blast had psychological as well as physical consequences, no one is really sure where *physical* radiation effects end and *psychological* radiation effects begin.

(b) Psychic Death Immersion: Death Guilt, Psychic Numbing, Impaired Mourning, Counterfeit Nurturance and Paranoia

The impact of the atomic bomb upon the bodies of the hibakusha is equalled by the impact it has had upon their psyches. The general psychological impact of being immersed in death and having one's life thrust suddenly into death-dominated life has been discussed. There are, however, particular amplifications of this general condition that need to be examined. To do this, I shall discuss the experience of the hibakusha from Hiroshima in connection with that of the concentration camp survivors of the Nazi Holocaust. There are two reasons for this; first, because what the Holocaust meant in terms of the possible extermination of a *particular people*, Hiroshima portends for the entire race as a *particular species*; i.e., Hiroshima and the Holocaust form a complementarity of macro and micro; and secondly, survivors of both events were forced to come to grips with an overwhelming immersion in death of unparalleled intensity.

This is not to say that Hiroshima and Holocaust survivors are the same. The survivors of Hiroshima are unique in three ways: first, because of the suddenness and totality of their death-immersion; secondly, because of the permanent presence of this death-immersion because of the long-term effects of radiation upon the bodies of both the living and those not yet born; and thirdly, because of the continuing relationship of Hiroshima with the present global possibilities of nuclear holocaust. Nazi concentration camp survivors, in contrast, underwent an experience less directly associated with the death anxiety of the contemporary global experience but were involved in a much longer period of humiliation and terror and were victimized by more generalized psychic and physical assaults. Concentration camp survivors are much more likely to retain diffuse and severe psychic impairment, therefore, while in hibakusha the death imagery has a tendency to be more exclusively predominant. Perhaps the most important difference between the two groups of survivors is that while release from the concentration camps meant release from the immediate encounter with death, there is no release from the effects of radiation. While concentration camp survivors carried with them psychological and perhaps physical impairments, therefore, the people of Hiroshima were subjected to a special terror made possible because of a scientifically constructed radiation exposure which has held them into a life-long sense of vulnerability to the 6 August experience. As one hibakusha put it: 'Auschwitz shows us how cruel man can be to man, an example of extreme human cruelty – but Hiroshima shows us how cruel man can be through science, a new dimension of cruelty.'[156]

With these distinctions in mind, one can still describe both experiences as that of being involved in what Lifton terms a 'death spell'.[157] This comes from a sense of having literally entered the realm of death itself and yet having returned from it, a dynamic which gives the experience its lasting and ubiquitous power. The death spell is a perennial reminder to the survivor that s/he has 'touched death' and has returned; a reminder of death, then, that acts as a reminder of survival as well.

The death spell can be associated with prolonged grief and mourning. Early symptoms include a preoccupation with images of the dead, guilt, bodily complaints, various erratic and anti-social reactions, which if not integrated can become chronic and seriously disruptive of normal patterns of conduct.[158]

The survivor mourns for both lost loved ones, the anonymous dead, and for the loss of inanimate objects and lost symbols. In sum, the survivor mourns for the loss of her or his former self, for what s/he was like prior to her or his death immersion. What has been taken from the survivor is something precious to the living – the innocence from death, particularly grotesquely demeaning death. Elie Wiesel, therefore, tells how one Holocaust survivor rushes back to tell the unwarned Jews with the words, 'I wanted to come back . . . to tell you the story of my death.'[159]

As Karl Stern explains it, the death spell involves 'dying with' those already slain and impels the survivors to feel their subsequent life and power as somehow connected with the event which took some and spared others.[160] From the death event onwards, therefore, the survivors are bound to the dead as well as to their own grief for the dead by an inability to recapture the pristine state they lived in 'before the fall'.

Paradoxically, the life of grief the survivors are thrust into is generally tainted by various kinds of impaired mourning; that is, a general inability to perform the 'work of mourning'. The Hiroshima survivors in particular were deprived of the ability to mourn because they had no way to *prepare* themselves or the dying ones for what befell them; and neither the Hiroshima nor Holocaust survivors were able to cope with the *totality* of the loss, both groups being simply overwhelmed by the magnitude of what they underwent. It is Lifton's impression that it was the inability to absorb the atomic bombing experience itself that constituted the basis for the hibakusha's lifelong state of grief; that is, that their sudden and absolute shift from ordinary experience to an overwhelming encounter with death is more because of the magnitude of the event rather than their lack of preparation for it.[161] Researchers of Holocaust victims have stressed a similar phenomenon, that of the 'missing grave', which expressed for them an impairment to mourning and a cause of later psychiatric difficulty. What is really at issue in both cases is that of the 'missing dead', which Lifton explains as 'the survivor's sense that the bodies – the human remains – around which he might ordinarily organize rituals of mourning, abruptly disappeared into smoke or nothingness'.[162] The impaired mourning, then, is due to two factors: first, particularly in the case of Hiroshima, there was no time for 'anticipatory mourning' which allows the living to prepare themselves for the departure of the loved one; and secondly, there was no allowance for proper burial or ritualistic grief and reformulation because, in the case of the Holocaust, Jews saw their friends, family members and fellow nationals enter the gas chambers, never to return; and in the case of Hiroshima, the bomb literally vapourized thousands, leaving the rest killed in the initial blast to be burned in mass cremations carried out by the army.

In terms of the 'end-of-the-world' imagery often expressed, similar principles apply. It can be said that those surviving the atomic bomb lived out in psychic and physical actuality experiences generally associated with psychotic delusion.[163] There is an interesting inversion here, however, for while the 'world destruction' delusions of the psychotic reflects an impaired relationship to the world in which s/he *projects* upon the world her or his own *inner* sense of 'psychic death', the survivors of both Hiroshima and the Holocaust reversed the process. For them, it was the devastating *external* experience of near total annihilation that connected with related tendencies in their inner lives. This is to say that the external experience merged with mental images already in the psyche which signify separation and helplessness, stasis and annihilation. The survivor's imposed image of world destruction is a symbolic reactivation of that sense of psychic death which everyone knows at some level, although it becomes a dominant motif in the lives of Hiroshima and Holocaust survivors not because of personal delusion but because of the enormity and grotesqueness of the external reality they were forced into.

It is this coming together in complementarity of both inner symbols of destruction and death with the external experience of destruction and death that makes the death encounter an indelible one, one which, to recall Arrupe, 'engraved' the mind, creating the ambiguous yet profoundly powerful image of 'ultimate death' and 'ultimate separation'. *This coming together of inner images of death and outer experiences of virtual annihilation produces such a radicalization of normal reality that the survivor's mental economy undergoes a permanent fusion of the two, the consequence of which is a 'psychic mutation'.*[164]

Rather than being a totally tragic event, however, this psychic mutation, in concentration camp inmates particularly, was necessary for their very survival. Niederland notes that 'there is reason to believe that a person who fully adhered to all the ethical and moral standards of conduct of civilian life on entering the camp in the morning, would have been dead by nightfall'.[165]

This near-complete ethical reversal was perhaps more pronounced in Holocaust survivors than for hibakusha because the Holocaust victims were forced somehow to cope with living in the face of extreme brutalization. The psychic mutation of the hibakusha, while not producing ethical reversals, has nevertheless been equally intense because the 'invisible contamination' of radioactivity continues to haunt them. The fusion of outer and inner continues to the present day with cancer, leukemia, and the genetic abnormalities among their offspring. In both cases, the driving force behind it and therefore the origin of the psychic mutation is the threat of death.

This intrusion of death into the cyclic pattern of the living is of profound import when one calls to mind Freud's claim that 'although his whole being revolted against the admission', it was only through the death of a loved one or friend that one is 'forced to learn that one can die, too, oneself'.[166] Lifton amplifies the point by emphasizing that the death anxiety of the Hiroshima and Holocaust survivors was not merely with dying itself but with premature death, grotesque death and ensuing unfulfilled life. The devastation of the death immersion the survivors underwent was as impactive as it was because the cataclysm had no reasonable relationship to either normal life-spans and dying or to any sense of historical justice or retribution. To any question of 'Why?' there was no acceptable 'Because. . . .' The only thing to be experienced, therefore, was a fundamental disruption in the general continuity of human existence and the impaired psychic and physical consequences of having been involved in such a disruption.

Inseparable from this disruption and the aftermath of death-dominated life is the survivors' struggle with guilt. From the immediate experience onwards the survivors were tormented by a fundamental ambivalence: they embraced the dead, paying homage and performing rituals in their memory as best they could; but they also pushed them away, considering them tainted and threatening, dangerous and unclean. This resentment towards the dead, despite an ongoing reverence for them, is strengthened by whatever difficulties the survivor has had in envisioning continuity with them. Quite simply, *the dead are resented for depriving the survivors of their sense of immortality*. This dimension of 'homelessness', of knowing experientially what Freud asserts most people refuse to believe, leads survivors to spend the remainder of their lives struggling to equalize their relationships to the dead in the face of resentments they can neither fully integrate nor express. The survivors resent the dead, therefore, for compelling them to realize, generally grotesquely, that they, too, will die.

This resentment is borne of identification with the dead. Recalling Wiesel's phrase in *Night*, that 'in every stiffened corpse I saw myself',[167] one can say that survivors simultaneously feel themselves to be that 'stiffened corpse', condemn themselves for not being it themselves and condemn themselves even more for being relieved that the corpse is really the other person's and not their own. It is just this process of identification which creates guilt, an ambivalence which leaves the survivors with their own intrapsychic version of what Lifton terms a 'wound in the order of being'.[168]

Identification guilt such as this is based upon a tendency

inherent in all people to judge themselves through the eyes of others. The significance of this is demonstrated by the internalized image many Hiroshima survivors had of being confronted by the accusing eyes of the dead. For one history professor, these eyes of the dead were the most powerful memory he had of the entire atomic bomb experience:

> I went to look for my family. Somehow I became a pitiless person, because if I had pity, I would not have been able to walk through the city, to walk over those dead bodies. The most impressive thing was the expression in people's eyes – bodies badly injured which had turned black – their eyes looking for someone to come and help them. They looked at me and knew that I was stronger than they. . . . I was looking for my family and looking carefully at everyone I met to see if he or she was a family member – but the eyes – the emptiness – the helpless expression – were something I will never forget. . . . There were hundreds of people who had seen me. . . . And I often had to go to the same place more than once. I would wish that the same family would not still be there. . . . I saw disappointment in their eyes. They looked at me with great expectation, staring right through me. It was very hard to be stared at by those eyes. . . .[169]

It is true that this pressure of identification is stressed more strongly in East Asian and other non-Western cultures where it is utilized frequently as a shame sanction than in the West where the emphasis tends to be upon internalized conscience and upon inner sin and sinfulness. Nevertheless, the distinction is a subtle one at best, for both Holocaust and Hiroshima survivors demonstrate that identification guilt can become thoroughly internalized and serve as a conscience for the person involved. Indeed, even the Western sense of sinfulness is to a large degree based upon a process of identification, whether with the religious book and/or community or with the image of Christ.

Both guilt and shame, then, are fundamentally an expression of human connectedness, or the lack ot it, and the eye symbolism recalled by so many Hiroshima survivors transcends the distinctions often made between the two. The fact that these eyes belonged in many cases to the anonymous dead suggests that guilt identification goes beyond those emotionally or otherwise close to the survivor to embrace all those involved in the catastrophe. Anonymous eyes in this regard have a particularly strong impact, almost as though representational of the 'all-seeing eye' of an unknown deity, or the evil eye of an equally demanding malevolent force.[170]

The cataclysmic experiences of Holocaust and Hiroshima survivors demonstrate that guilt is an immediate phenomenon when one participates in the disruption of the general human order and is in some way separated from normalcy. This is true whether the Western cultural idiom of sin and retribution is employed or whether the humiliation and social abandonment of the Orient is utilized. In what we are discussing, it is death, premature and gruesome death, that formulates the essence of that disruption and separation. In identifying so strongly with the dead, the survivors seek in some way both to atone for their participation in the disruption and to reconstitute a coherent order around that atonement.

It is the shifting patterns of identification formed by the survivor that produce the 'rhythms of guilt' observed in both Holocaust survivors and hibakusha. It is a rhythm that involves the survivor in a complex of images to which s/he identifies her or him self: from an image of purity modelled upon the dead to one of destruction and separation modelled upon the environment of the death immersion and possibly upon its instigators.

What is more, this identification guilt *radiates outward*. In Hiroshima this 'radiation' moved from the dead to the survivors to ordinary Japanese to the Americans who dropped the bomb to the rest of the world. This is to say, that the survivors feel guilty towards the dead, the rest of Japan feels guilty towards the survivors, the Americans feel guilty for having dropped the bomb in the first place, and the rest of the world feels guilty for having participated in some way with the event either by having sided with the Americans or simply knowing about it.

The dead – the eyes of the dead – form the core of the death immersion. Proceeding outwards, each group internalizes the suffering of that group one step closer to those compelling eyes, feeling both identity with the dead and alienation from those 'stiffened corpses' for reminding those still living of the mortality of human existence. However subtle this pattern is to be discerned at the periphery of its impact, its existence suggests that the guilt associated with identification provides an important basis for the ultimate symbolic connectedness of human behaviour and psychology. When one seriously considers the phenomenon of this death guilt complex radiating outward around the planet, one can then begin to appreciate the magnitude of the impact Hiroshima and the Holocaust have had on contemporary reality. When one considers further that both these horrors were products of human volition and behaviour, then one must include in the general guilt emanating outwards not only those *victimized* but those who *perpetrated* the deed. In both cases the guilt is engendered from the annihilation of the bonds of human identification and community through violently administered mass death on a premature and terrible scale.[171]

The tragedy about guilt in human affairs is that rather than seeking expiation or forgiveness, the overwhelming response seems to be that of the cessation of feeling. This is certainly true for Hiroshima survivors. After the initial death immersion, people reacted against the death anxiety and death guilt through a dynamic of *psychic closing off*, through *psychic numbing*. A similar pattern can be discerned among Holocaust survivors. People tend ultimately to reject rather than integrate the experience of death immersion.

Niederland argues that some psychic closing off is necessary for a person's survival under conditions as horrendously dislocative as the Holocaust was,[172] but taken beyond a certain degree psychic numbing destroyed the survivors' ability to grasp the deaths of others and the dangers to themselves. Wiesel's description of a group of Jews who had arrived at a concentration camp the previous day bears this out:

> Those absent no longer touched even the surface of our memories. We still spoke of them – 'Who knows what may have become of them?' – but we had little concern for their fate. We were incapable of thinking of anything at all. Our senses were blunted; everything was blurred as in a fog. It was no longer possible to grasp anything. The instincts of self-preservation, of self-defence, of pride, had all deserted us. In one ultimate moment of lucidity, it seemed to me that we were damned souls wandering in the half-world, souls condemned to wander through space till the generations of man came to an end, seeking their redemption, seeking oblivion – without hope of finding it.[173]

What can be seen from this depiction is that the Jewish survivors, 'damned souls wandering in the half-world', were experiencing the same sense of death-in-life encountered by the hibakusha, a state of such radically devastated existence that they no longer felt related to either the activities or moral standards of the life process. The insidious impact of psychic numbing on them was that of blunting their senses, blurring their thinking, dulling their instincts, until any sense of interpersonal caring was cut off.

Even in situations in which there is an attempt to care, to reach out to establish a genuine human relationship, the affect of psychic numbing is that of causing what Sandor Rado calls 'miscarried repair'.[174] Similar to the physical process which originates in the body's attempts to protect itself from toxic substances but which then transforms itself into a noxious pathological force, psychic closing off begins as a legitimate defence against an over-exposure to death but ends up overwhelming the person with death imagery. In Hiroshima, this miscarried repair took the form of later bodily complaints, particularly of fatigue and the lack of vitality. While arising to a large degree from the residual effects of radiation, this same phenomenon has been observed in Holocaust survivors as well. Researchers who have studied both, as Lifton has, suggest that the unifying psychological factor is that of miscarried repair due to psychic numbing. It is a psychological response bound up with death guilt and the feeling that vitality is somehow immoral in the face of the magnitude of death encountered. What can be discerned rather is a pervasive tendency toward sluggish despair, a continual state of psychic numbing which includes diminished vitality, chronic depression and a constricted life style which keeps a tight lid over the mistrust, alienation and rage seething just below the surface.

This expression of death guilt via bodily complaints is in keeping with the recent scientific hypothesis concerning psychosomatic phenomena in general, that they represent a 'final common pathway', a dynamic, according to Elliot Luby, formed by an 'entrapment or immobilization in an interpersonal field which is affectively perceived as threatening to life or biological integrity'.[175] This is to say, that the survivor of severe death immersion becomes permanently 'entrapped' by what s/he understands to be a continuous threat of death and is unable to dispel or to express this entrapment in any way other than the languages of the body. In this sense the bodily complaints of the survivors of Hiroshima and the Holocaust, to the degree their ailments were not physically caused, are a psychosomatic perpetuation of their original 'entrapment' at the time of the death immersion. The entrapment disturbs what J. Bastiaans calls 'the existing psychosomatic homeostasis' with a resulting pattern of 'pronounced psychosomatic symptoms' and a generally 'neurastenic syndrome', meaning nervous debility which includes weakness or exhaustion of the nervous system generally and the inadequate functioning of virtually any organ in the body.[176]

As in the case of guilt, an outward 'radiation' of psychic closing-off has been observed. The nearer one has been to the dead, particularly to the perpetration of mass death, the greater the energy to achieve the psychic numbing. The nearer to the dead, too, the greater the continuing struggle with guilt and the more likelihood of prolonged patterns of the psychic numbing permeating all the sectors of the survivor's life. Conversely, the farther away from the event, the easier the task of psychic numbing becomes. For those non-Japanese there may be an almost total emotional severance from the Hiroshima experience. What I am suggesting here is not different reactions as such but a continuum of different intensities to the same pattern of psychic guilt and numbing to an event which, if allowed to, would engulf the person in the imagery of death. At the centre as at the periphery, there always remains the potentiality for a re-opening of the psyche to the impact of the death immersion through either will or circumstance, something that will be discussed shortly, but the point being made here is that the response to guilt on the part of the vast majority of people and death event survivors is that of psychic numbing. In this numbing it is important to note that although death guilt is a fundamental factor, it is not death alone which induces the numbing but the *relationship* of death to the survivor's symbolization of life.

When the larger issues surrounding psychic numbing are examined, it emerges as perhaps the critical element in not only the general neglect of the human impact of the atomic bombing of Hiroshima but serves to explain why so little attempt has been made seriously to disarm the planet of weapons which dwarf the bomb used against Hiroshima and threaten the entire planet with annihilation.

A grotesque example of this can be taken from the Nazi doctors and medical researchers who used Jewish prisoners for brutal medical experiments. To the question of how a doctor, having taken the Hippocratic oath, could perform these activities, Betleheim replies: 'By taking pride in his professional skills, irrespective of what purpose they were used for.'[177] The victimizers as well as the victims respond to the guilt of the death-in-life situation in which they find themselves by psychically closing themselves off, thus enabling them to cope, in the case of the victims, and to continue the torture, in the case of the victimizers. One would think that when confronted with a situation that could mean the death of both parties, as the magnitude of nuclear weaponry today around the planet means, there would be pause to attempt a psychic opening up that would allow healing and the re-establishment of human relationship. Yet despite the spectre of 'double suicide', the testing and threatened use of nuclear weapons continues through a combination of technical-professional focus reminiscent of the Nazi doctors and perceived ideological imperatives that preclude emotional and humanitarian perceptions of what these weapons really do.

Psychic numbing is quite simply a focus for the constant paradoxes that confront us concerning the affirmation of life and the denial of death. It is both a way of maintaining life when confronted with unmanageable death immersion and it threatens, if uncontrolled, to dissipate the very vitality of the life being preserved. Since the technology currently available is capable of plunging every city into the Hiroshima experience, this aspect of the hibakushas' struggle – psychic numbing – is one that must be addressed, for it is what allows the arms race to continue that will in time, if unchecked, engulf more of humanity in nuclear holocaust.

A final theme which has been discerned as dominant in survivors' personal relationships and general world views is that of *counterfeit nurturance*. Lifton emphasizes here how feelings of personal need subsequent to the death immersion experience are combined with a great sensitivity to any reminder of physical or psychological weakness.[178] Psychic numbing, on the one hand, limits personal autonomy and cuts off potentially enriching relationships; on the other hand, any help offered threatens to confirm not only the weaknesses the psychic numbing is attempting to ignore but even more fundamental forms of devitalization.

Complicating this ambivalence toward nurturance is the

inevitable identification the survivors make with the death-dealing force itself. Among many of the hibakusha, there was a specific tendency to identify not only with the Americans but with the atomic bomb. Dr Hachiya, for example, describes his hospital patients as suddenly revitalized by the rumour that Japan had atomic bombed several cities along the west coast of the US 'Everyone became cheerful and bright,' he recalls, 'those who had been hurt the most were the happiest.'[179] For others, identification with the bomb was related to being in awe of its power, one grocer remembering that he felt '. . . only the greatness of the bomb'.[180] Still other hibakusha expressed almost a pride in having experienced the world's most advanced weapon. One scholar, who died shortly after the attack, was quoted as saying: 'The Americans are a great people, because anyone who makes such a terrible weapon must have greatness in them.'[181]

Even more striking was the unconscious identification on the part of Jewish concentration camp inmates with not only the ideology but even the specific mannerisms of the Nazi guards.[182] Although the general point argued is that power is in the final analysis power over death, the point being made here is more limited: namely, that the very formulation of a survivor's identity is built up around the victimization s/he feels at the hands of the force that can snuff out her or his life at will. It is an identification that forms a psychic bonding within the survivor to that force; it is an identification of magnetic relationship in which, as Lifton explains, 'the survivor feels drawn into permanent union with the force that killed so many others around him. His death-guilt is intensified, as is his sense that his own life is counterfeit.'[183]

It is within this context that one can begin to understand the more or less permanent 'victim consciousness' to which so many Hiroshima and Holocaust survivors have been susceptible. It is a consciousness born of an experience in which their entire lives were immersed in death, so much so that their very psyches became mutated, engraved by death-dominated life. Tormented by guilt, identification with the dead, most seek to eradicate the entire experience through the process of psychic numbing. This repression only serves to push the feelings deeper, and they will in time surface in psychosomatic disorders of mind and body. Nurturance is sought, relationships started, but the mistrust, alienation and guilt at even being alive before the staring eyes of the dead make much of the nurturance counterfeit. And so the survivors identify with the force that brought them into the death immersion, seeking solace through 'permanent union' with the force that killed so many others around them. And yet this only intensifies the guilt and the sense that their lives are counterfeit.

At its most extreme, victim-consciousness results in *paranoia*, a dimension of the survivors' response particularly noticeable in their world-destruction imagery. This paranoia is of course due to a major degree to the imprint of the actual brutalization they incurred, but of great importance also is the later struggle with rage over having been so devastatingly vulnerable and helpless. This is a pattern similar to that of counterfeit nurturance in that it is the survivors' way of expressing the feeling that they are still being victimized. It is also a desperate attempt to express a sense of vitality. It involves the mechanism of projection, a focus of attention upon the 'enemies' around the survivor and can lead to a superior form of adaptation to that of the Müselmanner in the concentration camps, although full-blown and disruptive paranoia was a virtual guarantee of death in the original death-immersion experience. Similarly, in Hiroshima, the anger and paranoia ensuing from the atomic bombing could in many cases protect a survivor from being overwhelmed by the psychic numbing, although like psychic numbing, it proved difficult to control and integrate, generally resulting in psychic impairment.

What emerges from the above discussion is a generally recognized sequence: there is first the actual death immersion and the experience of the 'end of the world', meaning that normal life has been suddenly transformed into death-dominated life; what follows is acute inactivation, residual death anxiety and death guilt and psychic numbing. Then the survivor is overwhelmed by a sense of impaired existence and the intense feelings that the nurturance s/he is receiving is counterfeit; the reaction to this is the focussing of rage to express simultaneously ultimate retaliation and active power over death; the final phase is a return, in delusional form, to the death immersion which started the cycle, a return which continues with the survivors as paranoia that their vulnerability will be held against them in whatever situation or relationship they find themselves.

Lifton suggests that this survivor paranoia can serve as a model for all paranoia, for paranoia considered generically is related to disturbed death imagery and therefore represents a struggle to achieve a magical form of vitality and power over death. It is at the same time an extreme form of suspicion, of counterfeit nurturance, in which the very help needed is perceived as deadly and something to be feared and countered.[184]

The question must be raised here as to whether there is any legitimate basis for the threatening and dark imagery that surrounds the survivors. It is a question that raises the problem of what has been termed 'survivor hubris', a tendency totally to embrace the survivors' special knowledge of death to the point of making that knowledge the central referent for their existence. It is a dynamic making sacred of the fact of death and the corollary awareness that they survived its onslaught. The effort to reinforce this energizing but fragile sense of power over death can often result in an actual addiction to or craving for the process of survival itself.

Death and destruction have always been the foci of fascination; it is a fascination composed of a denial of one's own death coupled with a fantasy of being what may be called an 'eternal survivor'. This is of course Freud's famous dictum, that 'at bottom no one believes in his own death . . . (and) in the unconscious every one of us is convinced of his own immortality'.[185]

Within this understanding, it is clear why Elias Canetti refers to survival as 'the moment of power', stressing that once tasted it can become 'a dangerous and insatiable passion'. He concludes that paranoia 'is an illness of power in the most literal sense of the words'.[186]

Paranoia, then, particularly among Hiroshima and Holocaust survivors, is involved with an intense struggle on their part to achieve a power over death. Because it is bound up with counterfeit nurturance, psychic numbing and guilt, however, it generally involves the survivor in an identification with the evil force that brought about the death immersion in the first place; so rather than paranoia enabling the person to overcome death, it merely pulls the psyche deeper into the morass of death imagery, yielding a world-view built around an absolute distinction between 'we who have made it through' and 'you who have not' and impulsive actions which endanger others.

Identification with the power that brought about the death immersion is the ultimate consequence of a failure to come to grips in a meaningful way with the guilt resulting from being alive among the dead. It is the result of psychically closing off rather than opening up to the enormity of the devastation done, a consequence of refusing to look into the staring eyes of the dead and dying and integrate them into one's own sense of pain.

As pointed out previously, this is a characteristic not limited to a few; it is a psychic response that involves all of us, particularly the nation that dropped the atomic bomb. What I attempted to convey in the dicussion of the milieu of war in which the decision to drop the bomb was made was this point, that the Americans and Western Allies in general, Britain most particularly, became fixated on the bomb, seeing it as not only the solution to the military campaign against the Japanese but as the solution to their problems with the Soviet Union as well. What we will see in the next section is how this identification with the bomb infected the entire matrix of American policy in the post-war world and has caused both Soviets and Americans alike to pour countless billions of dollars into a spiralling arms race that has yet to stop.

This identification is inextricably linked with two things: first, a process of *psychic numbing* rather than opening up to the deed done, a refusal, in other words, to come to grips with guilt; and secondly, identification with death bound up with *paranoia*, leading to a dichotomized world-view of 'us' and 'them' and destructive behaviour. This is the root of the counterfeit nurturance of the hibakusha and Holocaust survivor; it is the root of the current nuclear policies of the superpowers.

Perhaps if this negative formulation could be summarized in a word it would be that of the theme of 'nothingness'. Many hibakusha specifically stated that this was the word. One authoress, a Ms Ota, observed that 'the destruction returned everything to nothingness'.[187] A history professor utilized this theme as well, asserting that

> The story of Noah's ark is more than a myth to me. Except for a few humans and animals, it is a study of everything becoming nothing. Maybe this will happen again – everything disappearing and becoming nothing except for a very few. . . . If we continue to make and use more powerful bombs, there may only be a few people left . . . chosen by chance. . . . As for myself, I go the way of nothingness.[188]

What this meant in the actual lives of many hibakusha is brought out by a female poet in describing the life of her cousin:

> Her mother and father and all of her close relatives were killed by the A-bomb. She became very upset . . . and having lost everything, had to go from the house of one relative to another. During her wandering . . . she said she wished others would be made to go through such an experience so they would understand what she had gone through. . . . She had a 'crooked heart' at that time, so her relatives didn't like her very much. . . . When she expressed such feelings . . . her relatives repeated them to others because they were scared and thought she was an awful girl. . . .[189]

Indeed, so intense did this negative formulation and feeling of nothingness become for one woman that she said

> If it should ever happen again . . . I would die at once – at that very moment. This is all that a person like me can feel. Being already an old woman, if it should be used again, whatever may happen, I would rather die.[190]

This 'way of nothingness' in fact has close associations with a special version of psychological non-resistance which is related to Oriental psychology and which paradoxically led many hibakusha towards positive formulations in the end, something to be discussed presently, but for most 'the way of nothingness' led to nihilism and paranoic malajustment.

The way it gives insight into the special version of psychological non-resistance related to Oriental psychology is brought out by a history professor's comment, that 'Orientals tend to adapt themselves to their environment . . . while Westerners try to conquer their environment. . . .'[191]

A young writer put the distinction another way: 'At the centre of Western civilization is the idea of man. Man is less important in the Oriental Way.'[192] This attitudinal distinction has particular consequence on the subject of the atomic bomb:

> The Japanese Air Force attacked Pearl Harbour because they were interested in exhibiting their power by destroying the enemy's military forces. But when the Americans dropped the A-bomb, they were interested in exhibiting their military power through the number of people killed. At Pearl Harbour man remained man; at Hiroshima man was reduced to numbers.

While the Oriental perspective reduces humans to being

component parts of the larger whole, therefore, it paradoxically gives them a higher, more humane standing. The Western perspective distinguishes and elevates *homo sapiens* above nature and because of this has focussed all war technology, epitomized finally by the atomic bomb, on complete human destruction. What this means for the young writer is that 'the A-bomb represents the termination of Western thought'.

This is a profound point, for it points to the gravity of the Hiroshima event and the magnitude of the impact it has had upon the Western mind. We elevated our species above all things and now we have unleashed a technology that can destroy the only thing we hold to be sacred: ourselves. The tragedy is that rather than seriously reflecting upon this perspective of the Hiroshima event we have closed ourselves off from it, choosing instead 'the way of nothingness' exemplified by psychic numbing, counterfeit nurturance and a paranoia that now is able to turn every city on earth into another Hiroshima.

2. *Reformulation: Psychic Opening-up*

Because the threat of more Hiroshimas grows directly out of the counterfeit nurturance and paranoia stemming from psychic numbing and a refusal to come to terms with death guilt, experiences which enable us to break out of its hold are greatly valued. *Psychic opening-up*, if it can be done, is not only necessary to the resolution of the death immersion/death guilt complex; it can open up those experiences and utilize them for further affirmation of the life process so seriously threatened.

Lifton suggests three areas where positive formulation is essential: first, a sense of connection, of organic bonding with other people must be established; secondly, a sense of symbolic integrity, of cohesion and significance to life must be built up, a dynamic necessitating an integration and transcendence of the atomic bomb experience; and thirdly, a sense of movement must be felt, a sense of continuity and development in the continuous process of growing from the point of fixed identity to individuation.

What we have discussed thus far are those conflicts which impede or impair proper formulation. There are, however, avenues available to the psyche by which a formulation both beneficial and deepening can take place. This generally involves two predominant patterns: psychological non-resistance and the survivor's sense of mission.

(a) Psychological Non-Resistance

Psychological non-resistance involves a submission to larger forces one recognizes cannot be overcome. It suggests an encounter with ultimate mystery which must be absorbed rather than fought, assimilated instead of rejected. It is a psychological movement in which, to use Jungian terminology, the ego opens up to the self to receive whatever is there to be given. In the case of the hibakusha, this meant opening up to the encounter with the atomic bomb and absorbing into their persons the death it offered.

For many hibakusha, the atomic bomb experience was simply ineffable. They had no categories with which to interpret it and so reacted to it in a posture of awe and resignation. Such was the case, e.g., with Dr Hachiya, who remembers that 'in two days (after the bombing) I had become at home in this environment of chaos and despair. I felt lonely, but it was an animal loneliness. I became part of the darkness of the night.'[193] Because of this willingness to absorb rather than negate the experience, Dr Hachiya was to experience less residual negative psychological impact. Another hibakusha, a woman labourer, who like Dr Hachiya was badly injured by the blast, recalls that it was a sense of resignation which sustained her through the feelings of abandonment and later fears of 'A-bomb disease':

> . . . I am not too anxious about things. I do brood, sometimes but not to the extent of severe worry because I soon come to a feeling of resignation. . . . This helps me . . . and seeing that I have been getting along all right so far, I feel that it will be so in the future also. . . .[194]

This mode of non-resistance attempts to harmonize the individual with the event itself and represents a great deal more than mere passivity and hopelessness. What it involves is what Lifton terms 'a vision of ultimately indestructible human continuity',[195] *a perception predicated upon an inner confidence that one will be able to integrate whatever happens*, while looking beyond the event itself towards a reassertion of an inner imagery of the sense of connection, integrity and movement necessary to proper and positive formulation.

An insight into this process can be gained by examining the Japanese word for resignation, 'akiramu'. It connotes the idea of active encounter with powerful forces, not passive submission. Furthermore, the notion of illumination implied is not that of intellectual insight but rather the dynamic of giving these forces significant inner form.

Like the negative nihilistic forms of resignation, 'akiramu' also suggests the 'way of nothingness', but it suggests this way as a means by which the individual ego can blend with and be acted upon by the numinous forces and events surrounding it. It is a vision, then, in which there is infinite strife and change on one level and simultaneously a harmony of all the parts as well, even that part representational of death and destruction. In this resignation, one's survival is possible only through an enlargement of the dimensions of the physical and pshychological arena in which one is contextualized. It is a resignation, therefore, that breaks the boundaries of the individual ego, opening it to enlargement through absorption of the event or force encountered.

While this type of resignation is generally only achieved by people in the serenity of religious reflection, several hibakusha maintained this psychological non-resistance in the midst of the most devastating violence ever perpetrated on human beings by other human beings. They were not only able to master the guilt and anxiety of the situation but also to cope with the genuine 'end-of-the-world' imagery to come out of the devastation.

What is important is the religious component of formulation, particularly in Japan, influenced as it is by Buddhism, a religion which has cultivated virtually every form of resignation and psychological non-resistance. Many hibakusha directly equated their formulation through resignation with Buddhist belief and imagery, although even the most stalwart were shattered by the blast itself, as an elderly woman of the Konko sect demonstrates:

> At that moment we all became completely separate human beings. Seeing those wretched figures of people, I felt great pity. And having experienced such a terrible state of living hell, I thought, 'There is no God, no Buddha . . . there is no God, no help.'[196]

Her initial reaction, then, as with all directly impacted by the blast, was that of experiencing a shattering of the ego. Because she was shattered, 'experiencing such a terrible state of living hell', her concept of God was shattered as well. However, this particular woman did not revel in the death immersion but allowed herself to flow with the experience, deriving hope when different members of her family, though by no means all, began to reappear. She remembers that

> Even though in the middle of the painful experience . . . with no hope and everything burned down, I couldn't believe in the existence of God or Buddha . . . as time passed and the world became peaceful . . . I began to feel that we owe everything to God.

It was this sense of re-established equilibrium that allowed her to interpret the event retroactively in religious terms:

> I cannot attribute the fact that I was not killed to my own power. . . . I believe that God ordered me to put on a pair of sandals which just happened to be there. With all the glass about . . . I would not have been able to have walked . . . I attribute all of this to Konko-sama (the deity of her religious sect).

What is important to note in the experience of this woman is that once she was able to re-establish connection and meaning within her own religious imagery, she was able to restore it to her deity as well, indicating that from a psychological perspective one's inner sense of integration is directly impactive upon one's ability to appreciate the totality of God. The dynamic, then, is one in which the numinous at first shatters the ego and then the ego, as it begins to recoalesce, starts to re-integrate itself into the numinous which initially shattered it. Once re-integrated, the ego can then understand that even the shattering and the survival subsequent to the shattering are of numinous significance. One hibakusha is particularly clear on this:

> When I really think about it, I believe that my actually getting out was a matter of destiny. Whatever my own careful efforts, I feel there is a limit to what human beings themselves can do, and in regard to this moment I have the deepest respect for religion. . . .[197]

Resignation similar to this can be seen on the part of Christian hibakusha. Psychological non-resistance for Christians, however, stressed far more the active aspect of submission as a dynamic of the will to believe in an all-powerful God rather than the Buddhist notion of effortless harmony with the blast. One Korean woman, a fundamentalist Protestant, expressed a formulation that allowed her to come to terms with the loss of her children as well as a formulation through which she could come to terms with the human origins of the bomb in a world she believed God controls:

> Hiroshima was reduced to ashes by the power of human beings, by the A-bomb which was dropped. But still I came to believe in God's power – to believe that the world was made by God. . . . Even though we build big buildings, when the A-bomb is dropped, everything will be ashes. . . . Well, in the Bible – in the Old Testament when I read that, I thought that everything in the world is exactly that way – so everyone has to have faith.[198]

For her, the Hiroshima bombing signified punishment for human guilt analogous to the judgment on Sodom and Gomorrah. It was a punishment, however, perfectly consonant with an apocalyptic expression of God's will. Indeed, while many attempted religious formulations among hibakusha succeeded only in pointing to the direction the adherent *should* follow, this woman's ideal vision completely took command of her psychic life to the point where she could testify that because of the Hiroshima experience she was 'reborn'. This is not to say that conflicts did not remain; quite the contrary, particularly with regard to her lost children, she manifested the persistence of the most fundamental of A-bomb legacies: death guilt.

With the exception of the Korean woman, the research carried out by Lifton indicates that Christian formulations did not achieve any greater success than did the indigenous religious ones. In fact, only rarely did formulation of any kind contribute to the survivor actually gaining mastery of the atomic bomb experience; for most the death immersion was simply too devastating to be integrated. The enormity of the physical and symbolic breakdown, therefore, and the persistence of a particularly unabsorbable death anxiety and death guilt has permanently impaired most survivors.

What the dynamic of psychological non-resistance indicates, however, is that positive formulation is possible, even if successfully used by only a few. It means a posture of openness to the ultimately ineffable experience the atomic bombing was, a willingness to absorb the event rather than closing oneself off from it. It is a resignation that is not passive but active, offering the victim the means by which s/he can *re-connect* with the totality of being after the blast, thus avoiding the paranoia and ultimate death identification that impaired so many others. This is not to say that there was no psychic mutation. All involved in the Hiroshima bombing were mutated, for they were thrust from normal existence to death-dominated existence. The critical point lay in whether, as a response to this shattering experience, they would close their psyches off or open them up. Most closed them off, with the resulting counterfeit nurturance, paranoia and death identification; a few opened themselves up to the event and gained mastery of themselves and the ability to help others achieve meaning. For both, the path lay through 'the way of nothingness'; the choice they made, after finding themselves on this 'way', determined whether their destination lay in paranoia or meaning, nihilism or integration.

(b) A Sense of Mission

A sense of mission is similar to that of non-resistance, although while non-resistance seeks to absorb the impact with an ultimate reassertion of human continuity and/or divine providence, the sense of mission creatively utilizes the death guilt to help others, thus enabling the person involved to transcend the guilt. The two are in a sense complementary, for without non-resistance the hibakusha would be unable to absorb the losses involved, and without a sense of special mission, they would be unable to justify a meaningful continuation of life. A sense of mission would most generally follow in the life of a person who successfully absorbed the atomic bombing experience.

This complementarity is strikingly apparent among those hibakusha who emerged as leaders after the disaster. For all of them, a common theme was that of attempting to 'conquer death' through a demonstration of ways in which the devastating dislocation of life and death could be integrated into post-bomb existence. Through this effort they felt themselves making an effort to comprehend the fact of human mortality itself. This is a most difficult challenge, for it means not only coming to terms with the atomic bomb itself, but being able to combine one's personality with the larger idea of human community in such a way as to establish a rapport with what Freud terms the 'wishes' of those around one, either by reviving 'an old group of wishes' or by providing 'a new aim for their wishes'.[199]

As Erikson describes it, the challenge of mission is to 'increase the margin of man's inner freedom by introspective means applied to the very centre of his conflicts'.[200] The inner freedom gained by the hibakusha was through the path of non-resistance; the projection of this freedom to the centre of the conflict was the challenge of mission.

Perhaps the most outstanding example of this particular dynamic of formulation is that of a Hiroshima city official serving in a section concerned with the wartime distribution of food and provisions. Finding the City Office in flames when he arrived immediately after the bomb fell (he was 3,000 metres from the epicentre at the time of the explosion), he quickly set up a temporary headquarters in an adjacent, still-standing building. From there he worked, ate, slept and travelled throughout the city, helping to deal with the dead and caring for the living. Learning that the mayor was dead and seeing that the other officials, many higher than himself, were simply incapable of effective response, he simply and without hesitation took over the reins of power:

> They say I shouted at and directed the deputy mayor and other officials who were my superiors. I did not know I was doing this, as I was working like a man in a dream. . . . I cannot say how much I devoted myself to the work, but I do know that for the year I was doing it, I simply was not aware of whether I was living or not.[201]

His mission was to make Hiroshima 'a city of brightness', to develop 'a bright and forward-looking city population'; this because he was acutely aware that 'this experience should not be just confined to us. It is a great and significant experience – it should be shared with the world.'

This is not to say he was without enormous personal problems he was challenged to overcome. His personal experience of the psychological sufferings characteristic of hibakusha in general emerged to be intrinsic to his leadership. While carrying out the enormous task of locating goods and distributing them throughout the city, he had little time for contact with his immediate family, even when his own father-in-law lay dying. The father-in-law died alone, causing the official extraordinary feelings of grief and death guilt. Furthermore, he was 'terrified' with the other hibakusha at the prospect of being struck down by the 'invisible contamination' of radiation sickness, although rather than becoming morbid or neurotic about it, he compromised, continuing to direct the distribution procedures but instead of travelling all over, remaining in the central office.

This city official, then, lived out what Joseph Campbell defines as the classic pattern of the hero: answering the 'call', the 'summons' of the atomic bomb, which was simultaneously an 'awakening of the self' as well as a 'road of trials'. Even in the death of his father-in-law there is to be seen a symbolic 'atonement of the father' in which the official reasserted his paternal bond and then transcended it. Finally, there is to be seen both in himself and in his people 'the freedom to live'.[202]

Lifton suggests, in his own analysis of the man, that what was significant was his 'protean style of self process', meaning the ability he was able to demonstrate that at once allowed him to draw upon the old aspects of his tradition as a Japanese and resident of Hiroshima while simultaneously being able to embrace the new fact of the atomic bomb and being an hibakusha. This synthesis empowered him to make the type of boldly innovative decisions and strategems necessary for the city's survival. Furthermore, it permitted him to experience, both personally and publicly, the entire gamut of survivor conflicts, thus enabling all hibakusha to share in and identify with his individual death/rebirth symbolism. This same protean synthesis of old strengths interacting with new events contributed to his mediating skills by enabling him to experience all the complex nuances and convolutions of the A-bomb experience while making the types of decisions which sustained the renewed life of the community.

This process allowed both the man and the other hibakusha who identified with him to move beyond a preoccupation with either the past as it was before the bombing or the bomb itself and how it devastated them. It allowed an enlargement of their identity through absorption first of the bomb through the strength of their past and then a transcendence of the bomb through the articulation of a vision of 'a city of brightness'. Guilt, particularly over survival priority, did, of course, continue to plague the official, particularly as epitomized by the death of his father-in-law, but even if these guilt patterns weakened the overall heroic pattern that did not prevent him from attaining the heights of genuine heroism. Indeed, his example is one that illuminated the lives of many hibakusha and offers a model of the possibilities of positive formulation in the face of an overwhelming encounter with death.

3. Bearing Witness

Whatever the experience of the hibakusha, whether they survived nobly through positive formulation or were impaired grotesquely through an inability to formulate anything but a negative response, almost all have strong feelings about expressing their experience to others in order to convey the true nature of nuclear warfare.

While acutely aware of the limitations of language to express an experience that was totally engulfing as well as being equally sensitive to the dangers of misrepresentation and sensationalizing, hibakusha nevertheless continue to speak of their ordeal, reciting the event principally through autobiographical recall. Bearing witness through personal experience allows vicarious experiencing by the listeners, thus deepening the impact.

According to one hibakusha, he tells and retells his story so that those who continue to make more atomic bombs will 'think about it more from the inside, because if they did, they would feel differently'.[203] As one other hibakusha put it, 'I would like to tell the facts of the experience in a way that people will know about it not only with their minds but will feel it with their skin. . . .'[204] He is convinced that if the present world leaders who continue to stockpile more and bigger atomic warheads 'could have seen those people (at the time of the bomb) even once, they would feel that they should throw all their nuclear weapons to the bottom of the sea. . . .' The challenge of the hibakusha, then, is 'to help people understand the actual situation of human beings that day'.

This is to say, that in an attempt to appreciate the profound impact of the Hiroshima bomb, intellectual knowledge is not enough. The demand is that the observer immerse him or herself in atomic bomb exposure, allowing its 'invisible contamination' to permeate both body and psyche. The challenge is to penetrate through the psychic numbing we have built up to keep us from integrating and dealing with the death and survivor guilt caused by the 6 August 1945 experience. As Arrupe reminds us, 'the explosion of the first atomic bomb has become a para-historical phenomenon. It is not a memory, it is perpetual experience, outside history, which does not pass with the ticking of the clock.'

As another hibakusha puts it:

> In Hiroshima we have the fact of the A-bomb, and therefore we must stress the full horror of the A-bomb. Here is our responsibility and our emphasis. . . . In both Hiroshima and Japan in general the peace movement is based upon the fact of the A-bomb. . . . I think that the destruction of Hiroshima has important connection to the whole problem of human survival.[205]

This is to say, to quote Lifton, that 'organic knowledge of atomic bomb exposure . . . is Hiroshima's precious contribution to the world, an organizing principle for mankind's peace struggles'.[206]

Because the Hiroshima event encompassed the entire world in the psychological patterns of death guilt, psychic numbing and paranoia due to the fact of its ever increasing possibility of re-enactment, we are all survivors, we are all organically bound to its penetration of our psyches. Nevertheless, like most hibakusha, the world has cut itself off from this 'important connection to the whole problem of human survival' and has instead been caught up in the counterfeit nurturance and paranoia that such psychic cutting off engenders. That is the tragedy of the post-war period, that being drawn into the experience of death immersion, we rejected rather than absorbed the blast, and in our rejection have come to identify with death in such a way that our paranoia has propelled the planet into the Age of Overkill.

H. FROM TRUST TO TERROR: THE BEGINNING OF THE COLD WAR

According to Feis, the atomic bomb 'gave the Americans supreme power, of a kind, for a few years, and a false sense of security'.[207] They became so identified with the strength it represented and so mesmerized by the force it had demonstrated against Hiroshima and Nagasaki, he says, that they 'stood in awe of this super-human force and wanted to imprison it'. Herein was the paradox: the Americans were overwhelmed by the sense of power they derived from possession of the bomb yet they realized that somehow they should not wield this power, that they should attempt to 'imprison it'. And yet to do this placed them within another dilemma: 'They strove to devise and innovate a system of control which would safeguard the world against the demoniac energy compacted in the atom. But this effort was always haunted by fear that some other country – most probably the Soviet Union – would by deception procure the same weapon and use it to impose its will on others'.

On the one hand, therefore, the Americans knew that they must give up the very object that had given Truman his 'new feeling of confidence'; this because it was a weapon of such unparalleled destruction; on the other hand, they felt inhibited from doing so because of the paranoia that always accompanies death identification: they were fearful and mistrusting of the Russians that they would somehow 'by deception' develop the bomb on their own and use it against the Americans. Psychologically, then, the Americans felt compelled to disarm from the object of their identification and yet paralysed from doing so by a sense of paranoia and fear. The result of this clash between guilt demanding justice and fear compelling alienation was a situation in which, according to Edward Teller,

> when the war ended, the United States had no blueprint for pressing its atomic advantage. We stumbled forward with no concrete plan, no national policy outlining the conversion of our awesome weapon of war into a significant instrument for peace.[208]

Within this contradiction, Truman clung to the policy of secrecy. As will be recalled, when the knowledge of the successful testing of the atomic bomb had reached him in Potsdam and the first chance came to share this new breakthrough with the Russians, Truman, with the concurrence of Churchill, had waited a week and then only mentioned it in such oblique terms as to be incomprehensible to Stalin. This seemingly trivial refusal to be honest and open with someone who was then an ally set the tone for most of what was to happen later. Indeed, Feis remarks that

> . . . the evasions of Potsdam now seem clearly to forecast the devious direction that subsequent discussions about the control of atomic energy would take. The nations were to be frightened tenants in a house that at any time might be blown to pieces by a monstrous force rather than trust each other as caretakers.[209]

Truman's 6 August statement that accompanied the bombing of Hiroshima stated quite categorically that 'it is not intended to divulge the technical processes of production or all the military applications' of the atomic bomb.[210] Three days later, on 9 August, Truman declared that 'the atom bomb is too dangerous to be loose in a lawless world' and that therefore 'we must constitute ourselves trustees of this new force'.[211] On 15 August, Truman specifically directed the Secretaries of War, State and Navy, the Joint Chiefs of Staff and the Director of the Office of Scientific Research and Development

> . . . to take such steps as are necessary to prevent the release of any information in regard to the development, design or production of the atomic bomb; or in regard to its employment in military or naval warfare; except with the specific approval of the President in each instance.[212]

The first official reaction to Truman's refusal to share information came from the Soviet Union on 4 September, in the Soviet magazine *New Times*. The article emphasized the point that the American possession of the bomb did not solve the major political differences between the two countries. The Soviets attacked those in the American press and political arena who advocated the use of nuclear weapons to enforce American demands. The article furthermore made it clear that 'the fundamental principles are well known and henceforth it is simply a question of time before any country will be able to produce atomic bombs'.[213] For this reason, the article strongly advocated that an international body be entrusted control of atomic energy.

This advocacy of immediate international control found its most articulate and driving spokesperson on the American side in Secretary of War Stimson. While initially believing that any negotiations with the Soviets should be linked with an insistence that they modify their political system, he came to see after the actual bombing of Japan that negotiations for international control had to take precedence over any tangential wish on the part of the Americans that the Communist system be made more democratic.[214]

After his return from Potsdam, Stimson again consulted with the civilian and military personnel associated with him in the development of the bomb, asking them for further reflections on its deployment and status in American foreign policy. On 17 August, Robert Oppenheimer sent him statements reflecting the opinion of the Scientific Panel of the Interim Committee concerning contemplated legislation relating to atomic development.[215] The Panel stressed that there was no satisfactory defence against nuclear weapons; that it doubted whether either the maintenance or superiority of such weapons would assure national security; and that keeping the knowledge secret would only be counterproductive, serving only to engender mistrust and competition by others seeking that knowledge. In the light of these observations, the Panel strongly encouraged immediate movement on the part of the American government towards an international agreement to ban nuclear weapons and control atomic energy.

Secretary of State James Byrnes, however, held antithetical opinions to those of Oppenheimer and Stimson, and when he learned of the Panel's recommendations, he directed the Secretary of the Interim Committee, George Harrison, to advise Oppenheimer that it was simply not feasible to reach the type of agreement the Scientific Panel had in mind. His directive rather was that Oppenheimer as Director of the Los Alamos Laboratory vigorously continue research and development to improve the fledgling American atomic arsenal.

On 4 September just before Byrnes left for London for the first meeting of the Council of Foreign Ministers, Stimson met with him to discuss their differences. Byrnes remained unpersuaded of the imperative for meaningful international negotiations, however, and left for the London Conference continuing to believe that the American government could secure important strategic advantage by remaining the only country to have nuclear weapons. He had reached this conclusion because of his increasing distrust of Stalin and the Soviet Foreign Minister Molotov, believing them to have been 'perfidious' at Potsdam and 'conspiratorial' in the newly formed United Nations.[216]

After this failure to convince Byrnes, Stimson took the issue directly to President Truman, and on 12 September presented him with a memorandum expressing his views. This memo is of great importance as its incisiveness was equalled by a prophetic awareness of what would occur if meaningful negotiations did not take place:

> In many quarters (the atomic bomb) has been interpreted as a substantial offset to the growth of Russian influence on the Continent. We can be certain that the Soviet Government has sensed this tendency and the temptation will be strong for the Soviet political and military leaders to acquire this

weapon in the shortest possible time. . . . Accordingly, unless the Soviets are voluntarily invited into a partnership upon a basis of co-operation and trust, we are going to maintain the Anglo-Saxon bloc over against the Soviet in the possession of this weapon. Such a condition will almost certainly stimulate feverish activity on the part of the Soviet towards the development of this bomb in what will in effect be a secret armament race of a rather desperate character. There is evidence to indicate that such activity may have already commenced.

Whether Russia gets control of the necessary secrets of production in a minimum of say four years or a maximum of twenty years is not nearly as important to the world and civilization as to make sure that when they do get it they are willing and co-operative partners among peace loving nations of the world.

Those relations may be perhaps irretrievably embittered by the way in which we approach the solution of the bomb with Russia. For if we fail to approach them now and merely continue to negotiate with them, having this weapon rather ostentatiously on our hip, their suspicions and their distrust of our purposes and motives will increase. . . .

If the atomic bomb were merely another though more devastating military weapon to be assimilated into our pattern of international relations it would be one thing. We could then follow the old custom of secrecy and nationalistic military superiority relying on international caution to prescribe the future use of the weapon as we did with gas. But I think the bomb instead constitutes merely a first step in a new control by man over the forces of nature too revolutionary and dangerous to fit into the old concepts. I think it really caps the climax of the race between man's growing technical power for destructiveness and his psychological power of self-control and group control – his moral power. If so, our method of approach to the Russians is a question of the most vital importance in the evolution of human progress.[217]

Truman read the memo aloud paragraph by paragraph, indicating his agreement with its import as he went along. Stimson left the meeting, therefore, feeling that he had gained the President's approval. But the measures proposed by the Secretary called for bold initiative and trust in the midst of growing reports of Soviet expansionism in Germany, Poland and the satellite states, discord at the meeting of the Council of Foreign Ministers and Soviet belligerency in the Far East in taking territory in northern Asia from both China and Japan. Moreover, indications of the 'feverish' activity on the part of the Soviets that Stimson had warned of were already taking place. Commanders of the Czechoslovak Army were being compelled to turn over all German plans, parts, models and formulas regarding atomic energy, rocket weapons and radar to the occupying Soviet troops. The Soviets also moved into Jachimov and St Joachimstal and occupied the uranium factory there, the only place in Europe where uranium was being produced – even though such an occupation was in violation of the agreement that was in effect between Czechoslovakia and Russia.[218]

It was toward these immediate political realities that Truman directed his attention, not the long-term historical consequences of what the bomb meant. Indeed, he states specifically in his memoirs that long before the meeting with Stimson 'I had decided that the secret of the manufacture of the weapon would remain a secret with us', indicating that he was in agreement with Secretary Byrnes much more than with Secretary Stimson.[219]

Truman might have had the courage to seize the historical moment rather than merely react to political stimulii had more of his advisers agreed with Stimson. But they did not. On the last day of Stimson's tenure as Secretary of War, 21 September, a Cabinet meeting was held during which Stimson made an urgent appeal for adoption of the memo as policy. What ensued was the longest Cabinet meeting of the Truman administration. Stimson was supported by Acting Secretary of State Dean Acheson, Commerce Secretary Henry Wallace, Labour Secretary Schwellenback, Under Secretary of War Robert Patterson and Postmaster General Bob Harregan. He was vigorously opposed by Treasury Secretary Fred Vinson, Navy Secretary Forrestal, Attorney General Clark and Agriculture Secretary Anderson. The rest of the Cabinet made up the middle marginal position in between these two positions.[220]

In response to the intense debate Truman deferred decision and sought further advice from Vannevar Bush, Director of the Office of Research and Development, and from the Joint Chiefs of Staff. Dr Bush suggested that a proposal be made to the Soviets concerning the exchange of scientific information. He felt that keeping the dialogue, at least initially, at the level of scientific informational exchange could prove to be an effective way to control atomic proliferation.[221] The Joint Chiefs, however, recommended that the United States retain all existing atomic secrets as long as possible, at least until some of the fundamental international political problems were stabilized. Their advice to the President was that steps of a 'political nature should be promptly and vigorously pressed during the probably limited period of American monopoly' to restrict the release of any atomic information.[222]

Still undecided as to what to do, although in the meantime accepting the advice of Byrnes and the Joint Chiefs, Truman waited for the chance to discuss the matter with Prime Ministers King of Canada and Attlee of Britain when they met him in November 1945. The outcome of this meeting was a joint declaration, perhaps sincere in its intention to begin negotiations, but not genuinely clear enough of mistrust for the negotiations to be effective. The statement asserted that the signing parties realized that the world was 'faced with the terrible realities of the application of science to destruction' and committed themselves with 'firm resolve to work without reservation' towards 'creating conditions of mutual trust in which all peoples will be free to devote themselves to the arts of peace', all that was *in fact* done was to set up a Commission under the auspices of the newly formed and powerless United Nations to 'prepare recommendations for submission to the Organization'.[223]

The Commission assembled in June 1946. Both the Americans and the Soviets came with fixed objectives and tactics: the Americans seeking through the Baruch plan to eliminate nuclear weapons – provided that they could maintain their monopoly until they were sure no other nation could attain one, Truman stating: 'we should not . . . throw away our gun until we are sure the rest of the world can't arm against us';[224] the Russians through the Gromyko plan also seeking to eliminate nuclear weapons – provided that the American monopoly could be nullified while the Soviets strove to attain the weapon themselves.[225] Feis comments:

> The discussions in the United Nations were earnest. They bulged with professions of determination to dispel the menace of atomic weapons. But the proposals of the main progenitors of policy . . . never meshed; they were lofted

past each other by contrary winds of contrary purpose. Mistrust was so great, strategies so opposed, and social structures so divergent, that there was no middle ground on which the Western creators of atomic weapons and the Soviet Union, aspirant rival creator of them, could or did meet.[226]

In July 1949, three years later, the Commission finally adjourned so divided and antagonistic that the only resolution upon which the Soviets and the Americans could agree was that 'Further discussion would tend to harden these differences, and would serve no practicable or useful purpose'.[227]

By referring what are arguably the most important international negotiations ever held to the oblivion of the United Nations, Truman in effect sealed the fate of the subsequent history of atomic weaponry, even though Truman asserted that his motive had been to let the new organization 'prove itself'.[228] Secrecy on the part of the Americans continued throughout the negotiations, and the mutual mistrust grew until it had culminated in the 'Iron Curtain' in Europe and the McCarthy era in the US.

Within months of the adjournment of the deadlocked Commission, in August 1949, the Soviet Union successfully exploded an atom bomb of her own. The effort had been for the 'national security' of Mother Russia, even as the American effort four years before had been to protect the Allies. As the physicist Kurchatov, who directed the Soviet programme put it: 'The Soviet scientists knew that they had made a weapon for their own country to protect the peace of the world. . . . They had knocked the trump card out of the hands of the American atomic politicians.'[229] In October, in a speech to the UN General Assembly, the Russian Foreign Minister stated:

> It should not be forgotten that atomic bombs used by one side may be opposed by atomic bombs *and something else* from the other side, and then the utter collapse of all present-day calculations of certain conceited but short-witted people will become all too apparent.[230]

Molotov's words were prophetic, for the 'and something else' began to dominate the thinking of the American mind even as it began to dominate that of the Russians. Parity was not enough for either side; both sought advantage and superiority. Thus began the nuclear arms race or nuclear arms 'spiral', as Lerner puts it.[231] As soon as one side makes a breakthrough into the 'and something else' category, tension, resentment and fear are produced in the other side, which in turn serves to spur even more heroic efforts by the military/industrial complexes of both towards even bigger bombs and more accurate missiles and better communication systems, all within the context of attempting to tighten the respective alliance systems in play protecting each superpower, drawing in non-aligned nations if possible and disrupting the alliance system of the other where practicable.

The nuclear arms spiral is therefore a deadly complex of interacting factors which tend to heighten the tension and worsen the whole situation, producing what psychiatrists call a 'syndrome' a connecting web of rivalry in nuclear weapons, a warfare both hot and cold in the political arena and an atmosphere permeated by paranoia and mistrust. It is a syndrome that has been characterized by the escalation, by the idea of weapons generations, meaning the replacement of obsolete systems with new more lethally accurate and deadly systems, by threats, blackmail and xenophobia of the most provincial kind, and finally by a neurosis of secrecy and a spy-counter-spy mentality. What it has produced, says Feis, is 'a peace resting not on mutual trust but on mutual terror'.[232] And the terror has reached such a point that, according to former Secretary of State Henry Kissinger, the superpowers have more than enough nuclear weapons between them to 'annihilate humanity'.[233]

I. THE NUCLEAR ARMS SPIRAL

The American search for the 'and something other' was launched shortly after the Soviet detonation of the atomic bomb. While scientists had developed the Hiroshima and Nagasaki bombs on the principle of fission, it was thought possible to construct one on the principle of fusion, meaning that instead of *ex*ploding an atom one would *im*plode it.

On 29 October 1949, a mere month after the Soviet detonation, the General Advisory Committee of the US Atomic Energy Commission (AEC) met to give an opinion on the advisability of developing a thermonuclear bomb based on fusion. To develop such a weapon would be as much of an advancement over the Hiroshima-type fission bomb as the Hiroshima bomb had been over the tiny blockbusters of World War II. This time, however, the scientists were aware of the potentiality of what they were doing, and, headed by Oppenheimer – who had remarked after Hiroshima that 'Physicists have known sin' – the General Advisory Committee voted unanimously against any fusion bomb programme. Their report stated:

> We all hope that by one means or another the development of these weapons can be avoided. We are all reluctant to see the United States take the initiative in precipitating this development. We are all agreed that it would be wrong at the present moment to commit ourselves to an all-out effort towards its development. . . . In determining not to proceed to develop the Super bomb, we see a unique opportunity of providing by example some limitations on the totality of war and thus eliminating the fear and arousing the hopes of mankind.[234]

A minority report put the unanimous opinion in even stronger terms. Signed by Enrico Fermi and I. I. Rabi, it said:

> The fact that no limits exist to the destructiveness of this weapon makes its very existence and the knowledge of its construction a danger to humanity as a whole. It is necessarily an evil thing considered in any light. For these reasons, we believe it important for the President of the United States to tell the American public and the world that we think it is wrong on fundamental ethical principles to initiate the development of such a weapon.[235]

The forces within the Truman administration and the military that still believed that nuclear weaponry superiority could be translated into political advantage were overwhelming, however, despite the plea by the scientists. Among those insisting that research and development proceed were Secretary of State Byrnes, AEC (Atomic Energy Commission) Commissioner Straus, Senator McMahon, Chairperson of the Joint Congressional Committee for Atomic Energy, the Pentagon, and various scientists, the most vocal of whom was Edward Teller, eventually to head the hydrogen bomb development and earn the title 'father of the H-bomb'.

The catalyst for the decision to go ahead with the programme came in January 1950, when Klaus Fuchs, a scientist who had worked with the American team during the Manhattan Project to develop the atomic bomb, was arrested as a Communist spy.

According to Teller, Fuchs had been able to give the Soviets virtually complete knowledge of the American activities from as early as June 1944, which means that when Truman lied to Stalin at Potsdam about the bomb in July 1945, Stalin may well have known most of what Truman knew.[236]

Four days after Fuchs' arrest. Truman overrode the recommendation for restraint in the interests of peace by the General Advisory Committee and directed the AEC to proceed with its thermonuclear programme. On 1 November 1952, the US detonated a fusion bomb in the south Pacific; by August 1953, the Soviet Union had done the same.[237]

By November 1958, there had been detonated by both the US and the Soviet Union in their respective nuclear programmes more than 100 times as much explosive power as that dropped on Germany in all the years of World War II, the megatonnage being equivalent to the total body weight of the entire human population.

By 1960, John Kennedy observed that 'the world's nuclear stockpile contains, it is estimated, the equivalent of 30 billion tons of TNT – about ten tons for every human being on the globe'.[238] In that year alone, one US university accepted $40,000,000 in defence contracts, with forty other universities receiving $1,000,000.

And still the pace continued, incorporating more money, more scientists, more commitment on the part of the nations involved. By 1978, the US had enough nuclear explosive power to be able to destroy completely every Soviet city above 100,000 in population *forty one* times over; the Soviets could do the same to American cities above 100,000 *twenty three* times over. And still the spiral continues, until the multiplicity of weapons systems, delivery systems and satellite systems co-ordinating the entire complex of nuclear 'preparedness' has become truly staggering. The Stockholm International Peace Research Institute 1978 Yearbook, *World Armaments and Disarmaments*, lists five different nuclear weapons *categories*, twenty-one different nuclear weapons *types* with thirty-seven different maximum *ranges* and thirty-four different kiloton/megaton *yields, just for the nuclear weapons deployed in Europe* by the North Atlantic Treaty organization (NATO) and the Warsaw Pact Organization (WPO).[239] Moreover, these figures are merely for *tactical* nuclear weapons; the *strategic* nuclear weapons with higher yields and ranges, the US and the Soviet Union reserve exclusively for each other.

Perhaps the weapon system that symbolizes both the callousness of the nuclear arms spiral and its growing sophistication is the enhanced-radiation reduced-blast tactical nuclear weapon: the neutron bomb. The neutron bomb is the first nuclear battlefield weapon specifically designed to kill people rather than destroy military installations or property. While a building would be left intact, therefore, there would be 'almost immediate incapacitation of its occupants', according to an AEC document, meaning convulsions, intermittent stupor and a lack of muscle co-ordination that results in death, usually in a few hours.[240] Put in military language, as Alfred Starbird, Assistant Administrator of the Energy Research and Development Agency, does, 'You reduce the blast effect and get the kill radius you want. . . .'[241]

The leading advocates of this 'mini-nuke', as it is called, are the US generals commanding NATO. They have argued that the 7,000 tactical nuclear warheads now in Europe are 'old', have too large a yield and are therefore 'not credible'. With a lower yield and with a neutron blast that only kills people, not property, the neutron bomb, as the first generation of mini-nukes, offers a 'credible deterrent', the generals say – 'because they could be used without leading to automatic nuclear escalation'.[242]

In adopting use of the neutron bomb, according to the Stockholm International Peace Research Institute, the 'US doctrine has changed to make the use of nuclear weapons more likely in a tactical situation'.[243] It thus encourages the proliferation of nuclear weapons to countries which at present do not have them, as well as making nuclear war more 'thinkable' in the countries which already do. This is a danger even admitted by the White House in the Arms Control Impact Statement about the neutron bomb sent by the National Security Council to the US Senate: that the bomb may be perceived as a shift in US policy towards an acceptance of limited nuclear war.[244]

The neutron bomb is as insidious as it is because it epitomizes a reversal of sorts in the nuclear arms spiral. At the outset and throughout the 1950s and 1960s, the competition was that of making bigger bombs, longer range missiles, and missiles with multiple warheads. It was an attempt to *macrotize* the respective arsenals of the five countries involved: the US, the USSR, Britain, France and China. What emerged from this quantitative race was the doctrine of Mutual Assured Destruction (MAD), the notion that to go to war meant inflicting totally unacceptable damage on the enemy. Thus for most people war became unthinkable because it had grown to the proportions of the absurd. With the neutron bomb, however, the opposite concept of MAD comes into play: namely, *limited nuclear war*, the concept that if the nuclear weapons' arsenals can be *microtized* enough, they can be utilized on the battlefield without necessarily escalating the conflict into a mutual assured destruction scenario. James Schlesinger, when he was Secretary of Defence, along with former Secretary of State Kissinger and former President Nixon are perhaps most responsible for the articulation and adoption of this policy. Its effect has been, to quote the Stockholm Institute, that 'In both the USA and the USSR, groups who are thinking in terms of the feasibility of fighting a nuclear war may be gaining political influence.'[245]

What is occurring, then, is a gradual *blurring* of distinction between conventional weapons and nuclear weapons, for now, with the concept of mini-nukes, the tactical nuclear arms are being designed with kill powers equal to that of the conventional arms. This is producing a mentality in both camps that views the difference between nuclear and conventional arms not as *qualitative* but as merely *quantitative*, the nuclear weapons being perceived as a higher level of escalation along a sliding scale of 'appropriate' and 'credible' responses to any given situation.

Concurrent with the development of mini-nukes such as the neutron bomb are further sophistications in the strategic systems. This can be seen in President Carter's recent decision to deploy the MX missile to replace the older generation of intercontinental ballistic missiles, the Minuteman III missiles and the Titan missiles. Each Minuteman III carries three warheads, the Titans only one. The new MX system is projected to include 300 missiles, each one of which will have between ten and fourteen warheads with an explosive capacity of at least 200 kilotons (the Hiroshima bomb was 15 kilotons). The path of the MX warheads can be adjusted while in flight to elude enemy defences, a feature known as 'MARV' – Manoeuverable Reentry Vehicle.

The MX system is considered attractive to US defence planners for two other reasons. The first is that each warhead will not only be twice as powerful as the current Minuteman IIIs, but will have three times the accuracy: each MX warhead is capable of hitting within 100 feet of its target after travelling 8,000 miles.

With 300 missiles each carrying from ten to fourteen warheads, this means that the US will be capable of launching a force of between 3,000 and 4,200 warheads in a 'limited' first strike against the Soviet Union.

The second 'advantage' of the MX is that while extremely accurate and powerful in its offensive capacities, it will serve a defensive role as well. This is because the MX system will be deployed in silos throughout the US, according to what the US air force calls the 'shell game'. Each of the 300 missiles will be placed in underground trenches that are planned to be from six to twenty-five miles long. This will make it virtually impossible for the Soviet intelligence system to know precisely where any of them are. Such an underground track system is estimated by the White House to cover up to 10,000 square miles at a cost of about $5 million a mile. The missiles themselves are expected to have price tags of at least $100 million apiece.

Complementing the MX system is the new Cruise missile system: subsonic, pilotless airplanes between fourteen and twenty feet long that are capable of flying at treetop level and guiding themselves by a computer system that enables them to elude enemy radar and follow the terrain to their target. Like the MX, the Cruise missiles are capable of coming within 100 feet of their target after travelling thousands of miles. Unlike the MX, however, the Cruise missile is relatively cheap by defence standards, costing a mere $1 million per missile. Moreover, the Cruise missile can be launched from the ground, from airplanes, even from submarines, thus making detection virtually impossible. In 1979, President Carter ordered full scale development of the Cruise missile; the air force and navy are expected to order at least 5,000.

As if the neutron bombs, the MX missiles, the Cruise missiles, are not enough, the US military is also developing a new atomic submarine, the Trident, to replace the older atomic submarine version, the Poseidon. The Trident is over two football fields in length and over five stories high, making it more than twice the size of the Poseidons it is replacing. Besides size, the primary difference between the Trident and the older submarines is in the greater number, power and accuracy of its missiles. Each Trident will carry 50% more missiles than the Poseidons, and each missile will be twice as destructive. Each Trident will carry as many as twenty-four ballistic missiles, each with up to fourteen warheads capable of travelling 4,000 miles. This means that each Trident submarine will be able to deliver at least 192 city-destroying warheads.

The cost of the Tridents is expected to exceed $30,000,000,000.

Equally sophisticated as the mini-nukes and the newly designed strategic systems are the increasing complexity and accuracy of the guidance and communication systems which allow the warheads only to miss their targets by a few tens of metres, after travelling perhaps 10,000 miles up through the earth's atmosphere, and then down again towards the target. The newest guidance system to be announced by the Pentagon (in late May 1979) is that of a constellation of twenty-four navigating 'stars' in the form of satellites which have been specifically established to give unprecedented accuracy to US nuclear missiles. Called the Navstar Global Positioning System, it offers the US its latest in first-strike capability, assuring that any warhead it directs will be able to hit within ten metres of the target, thus giving Pentagon strategists confidence that they would be able to hit and crack Soviet missile silos and therefore destroy their missiles before they can be launched in retaliation. According to Lieutenant General Richard Henry, commander of the Air Force Space and Missile Systems Organization, the Navstar is 'so staggering that the strategic and tactical doctrine of our fighting forces will be rewritten'.[246]

Four Navstar test satellites are in orbit already, at a cost of $12,000,000 apiece. Like the twenty-four operational satellites that are to follow during the 1980s, they orbit at 10,900 feet and carry atomic clocks which the air force claims are accurate to within one second in 30,000 years. The cost of each clock: $200,000.

Although Navstar has serious implications for carrying the nuclear arms spiral into space, the US Secretary of Defence, Harold Brown, claims that the US effort is legitimated because the Russians are carrying on such satellite research as well, particularly in the area of killer satellites, which are capable of destroying other satellites while in space orbit. Such killer satellites, says Brown, 'pose a threat to all satellites' used by the US. With Navstar, however, the US has emerged 'comfortably ahead' in this dimension of the US-USSR competition.

It might seem that there would be a contradiction between this intense race for more kill-power, greater accuracy of delivery, and more complex communication and guidance systems, all in order to have greater 'credibility' of being able either to destroy the enemy before it can get its reactive missiles into the air or to be able to avoid any first strike and be able to retaliate with more punishment than the attacker, within the context of the Strategic Arms Limitation Talks (SALT). For while the nuclear arms spiral has continued unchecked, the two superpowers, as in the early days at the UN, are simultaneously talking about disarmament. While the UN conference adjourned in deadlock, however, the negotiations now have equalled the weapons systems in sophistication. Therefore, President Carter and Communist Party leader Brezhnev can now sit down in Vienna, as they did in early June 1979, sign the SALT II agreements, and yet still go home to continue working on their killer satellite systems as the Russians are doing, and their MX system, as the Americans are doing. This can be done within the context of 'meaningful' disarmament talks because the limits SALT II places upon the respective areas of nuclear competition are limits which in many cases have still *not been reached* by either side – thus allowing continued expansion of current programmes. Thus Secretary of Defence Brown can defend the announcement concerning the new MX missile system just prior to the signing of the SALT II accord by explaining to the public that the accord allows each side to develop an additional missile system. This 'creative ambiguity' in the SALT II treaty can be seen in the chart opposite.[247]

What emerges, then, is a curb in some systems with an allowance for growth in others. It is a treaty which has already been outdistanced by systems such as Navstar and killer satellites, neutron bombs and MX missiles. In effect, therefore, while capping the most lethal and oldest nuclear weapons, SALT II allows the arms race to continue in other areas. This is to say that after thirty-four years in the Age of Overkill there has been no serious attempt to begin meaningful disarmament of the nuclear weapons that threaten an entire planet.

J. FALLOUT

It is important to realize that during the entire time that the nuclear arms spiral has been going on, ordinary people have been having to pay for it, not only through their tax dollars but with their lives. Nuclear weapons did not only kill thousands in Hiroshima and Nagasaki; their continued testing has killed people in the United States, in Canada, in Europe, in the Soviet

A CEILING ON THE ARMS RACE

SALT II sets an over-all limit that will require the Soviet Union to scrap some of the older weapons in its current arsenal. Both the US and Russia can deploy more new missiles, but sub-limits restrict the most lethal weapons on both sides.

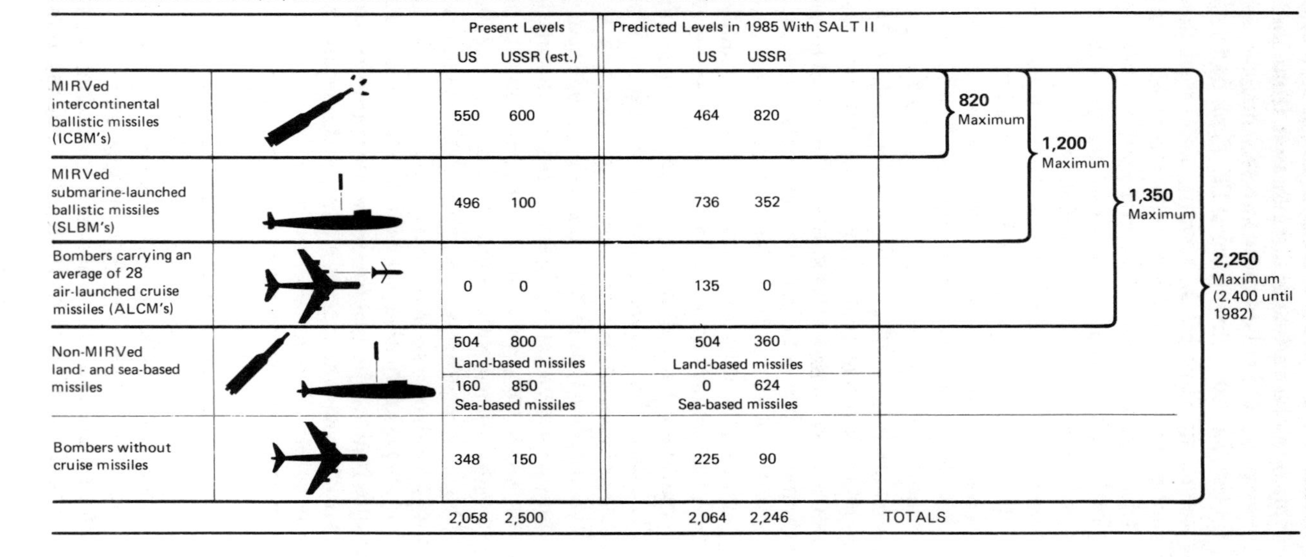

	Present Levels		Predicted Levels in 1985 With SALT II	
	US	USSR (est.)	US	USSR
MIRVed intercontinental ballistic missiles (ICBM's)	550	600	464	820
MIRVed submarine-launched ballistic missiles (SLBM's)	496	100	736	352
Bombers carrying an average of 28 air-launched cruise missiles (ALCM's)	0	0	135	0
Non-MIRVed land- and sea-based missiles: Land-based missiles	504	800	504	360
Non-MIRVed land- and sea-based missiles: Sea-based missiles	160	850	0	624
Bombers without cruise missiles	348	150	225	90
TOTALS	2,058	2,500	2,064	2,246

820 Maximum (ICBM's)

1,200 Maximum (ICBM's and SLBM's)

1,350 Maximum (ICBM's, SLBM's and bombers carrying ALCM's)

2,250 Maximum (2,400 until 1982) (all categories)

Union – people all around the globe. This has been done through the radioactivity released from the testing of nuclear weapons: it is called radioactive *fallout*.

On 22 April 1953, the AEC exploded a forty-three kiloton device at Yuca Flats, Nevada, as part of the Upshot-Knothole series of twelve nuclear tests that were carried out that year. Exploded at less than 100 metres above the desert floor, the blast formed a radioactive cloud that began drifting east. By 27 April it had drifted to within 200 kilometres of New York City, where it encountered a thunderstorm that rained down its radioactivity on the environment and populations below. This 'rainout' deposited between 35 and 70 curies of radioactivity per kilometre around the cities of Troy and Albany, New York. The fallout was noticed quite by accident by a professor conducting some experiments with his students in radiochemistry, whose geiger counters began to give readings of radioactivity far above normal. The professor called a colleague in the AEC Health and Safety Laboratory in New York City, and the AEC proceeded to conduct a study of radiation levels. The final report was classified and therefore not released to the public, the AEC merely stating that 'Fallout radioactivity is far below the level which could cause a detectable increase in mutations, or inheritable variations.'[248] The AEC announcement went on to assure people that any strontium 90 they might notice was the only possible hazard, but would arise only from 'the ingestion of bone splinters which might be intermingled with muscle tissue' of any of the cows they ate which had pastured on contaminated grass. Because there were no immediately noticeable effects from the fallout, this assertion was not challenged, and the incident was forgotten.

On 1 March 1954, on Bikini in the South Pacific, things went differently. The test involved the explosion of a hydrogen bomb which was expected to yield a force equivalent to about seven megatons. The actual blast produced twice this force. Again the resulting radioactive cloud began to drift, sweeping first over an American destroyer which responded by calling an emergency radiation drill which put all the crew below deck and sealed all the entrances.

On the islands of Uterik, Rongelap and Rongerik, some 160 kilometres away, however, no one knew about either the testing or about radiation drills. When the fallout reached there, the islanders, as at Hiroshima and Nagasaki, were unaware and totally exposed. Many sustained burns, spotty marks on the head, pigment changes on different parts of the body, scarring, and skin lesions. There were also cases of anorexia (appetite depression), nausea, vomiting, and abnormalities in the blood. The next sixteen years would see a massive increase in thyroid abnormalities, particularly among the children who had been less than ten years old at the time of exposure: seventeen out of nineteen developed a thyroid abnormality and two of them were dwarfed for life.

Passing beyond these people the cloud continued to drift eastward, passing over a Japanese fishing boat trawling for tuna well beyond the perimeters of the testing zone. All of the crew came down with radiation sickness; one was dead by 23 September.

The resulting worldwide outcry pressured the AEC to declassify some of its documents on nuclear weapons testing. One scientist to look at them, nuclear physicist Ralph Lapp, pointed out that these tests were being carried out under false assumptions. The AEC had been assuming that the fallout would be deposited uniformly over the entire globe and that it would remain in the upper layers of the earth's atmosphere for years, during which time the radioactive isotopes with shorter half-lives would decay away. What was actually happening, however, was that the fallout was coming down in a matter of months, if it went up into the higher atmosphere at all, and it was distributing itself primarily to the middle latitudes, where most of the human population is clustered. Lapp further pointed out that it was not how long it took some of the radioactive isotopes to decay that mattered but the peak concentrations in local areas when the fallout finally hit the ground.[249]

The controversy these revelations stirred caused further public outcry and the beginning of systematic studies concerning fallout. In 1957, the AEC Biological and Medical Advisory Committee concluded that contrary to previous AEC assertions, fallout from nuclear tests was in fact even causing genetic mutations. It projected between 2,500 and 13,000 major genetic defects per year in the global population as a result of the weapons testing up to 1956. Despite this data, the above-ground testing of nuclear weapons continued, indeed intensified. It was not until 1963 that the nations agreed to stop, deciding to continue their tests underground instead.

The effects of fallout from distant nuclear tests were first studied by a non-nuclear country in 1962 by the Ministry of Health in Alberta, Canada. Its survey included the years 1959 to 1962 and concluded that radioactive fallout causes sudden and statistically significant increases in the rate of congenital malformations, corresponding fairly closely with the rate of rainfall over a given region. What this meant specifically for the residents of Alberta was that because of the Soviet tests at Navaya Zemlya, north of the Arctic Circle in 1958, there was a 78% rise in birth defects in the following two years, the greatest increases being in the northern parts of Alberta, gradually lessening the further a population was away from the testing area. The study's most important discovery is that *each radioactive atom is some 10 million to 100 million times more toxic to developing embryos than a molecule of even something as devastating as thalidomide.*[250]

From the early 1960s onward, studies in this area were also carried out by Professor Ernest Sternglass, Professor of Radiation Physics at the University of Pittsburgh. What he found indicated that even the low-level radiation fallout that occurred in such places as Troy and Albany in 1953, produced marked increases in the leukemia rate among children. In 1969, at an AEC-sponsored symposium, Dr Sternglass delivered a paper stating that according to his studies some 400,000 infants less than one year old had probably died as a result of the nuclear fallout between 1950 and 1965. The eminent British radiobiologist, Dr Alice Stewart, was also at this symposium. It was her studies, published in 1958, that had first indicated that diagnostic X-rays given to pregnant mothers could produce a dramatic increase in the cancer rates of their children who were exposed while *in utero*. The dosages she studied were roughly equivalent to the dose of the fallout over Troy and Albany that Dr Sternglass had studied.[251]

The United Nations Scientific Committee on the Effects of Atomic Radiation has summarized the levels of radiation in the environment from all the tests by all countries involved as at the end of 1975. These summaries are given in terms of the total dose worldwide. The range is between 200 millirads in the bone-lining to 100 millirads to the gonad. For the middle latitudes, where most of the fallout has fallen, the rates are approximately 50% higher in each case. The highest dose from a single radionuclide is that from carbon-14, which remains toxic for well beyond its half-life of 5,600 years. The most deadly radioactive element, however, is plutonium-239, which according to Dr John Gofman, former Assistant Director of the AEC Lawrence Radiation Laboratory, is so lethal that between 116,000 to 1,000,000 cancer deaths will occur from its fallout in the United States alone.[252]

The most recent serious fallout incident is connected with the Chinese nuclear weapons testing. On 26 September 1976, they detonated a 200 kiloton device near Lop Nor in north-west China. The high altitude portion of the fallout reached the west coast of the United States on 1 October, passing in an easterly direction until it met a rainstorm along the eastern seaboard of the country. The radioactivity mixed with the rain, and a heavy washout of radiated water fell over a narrow strip of coast stretching from Delaware and New Jersey, across Connecticut, Rhode Island, Massachusetts, New Hampshire and Maine. Again Dr Sternglass studied the radiation levels, particularly the iodine-131 levels in the milk. Milk at certain dairies reached levels of 300 to 800 picocuries per litre, well over the 100 picocuries per litre set by the US Federal Radiation Council in 1961 as allowable for continuous consumption. Since this level had been raised in 1964, however, to an allowable dosage of 2,000 picocuries per litre, there was no mandatory requirement to take the milk off the market. Despite this federal refusal to acknowledge the dangers of fallout, Dr Sternglass found the following. In the areas of Massachusetts and Rhode Island where the milk was either produced by cows placed on stored feed or where it was imported, the infant mortality rate declined by 30%, in keeping with a national average of decline. In the areas where milk was taken directly from cows allowed to feed on the contaminated grass, however, the infant mortality rates increased from 13% to 60%.

What emerges from the studies of fallout, then, is the fact that radiation, whether it comes from the explosion of a bomb as in Hiroshima or in an invisible radiation cloud drifting in the atmosphere, is lethal to human life; indeed, it is reasonable to believe that as many people have been killed from fallout around the globe as were killed by the two bombs dropped on Japan. Although only the studies indicating this fact have been given here, a longer explanation of why and how this occurs will be presented in the following chapter. I only raise the issue now to indicate that the nuclear weapons experience is not something we can relegate to the people of Hiroshima and Nagasaki; it is an experience that involves us all.

This point can even be illustrated by the experience of the 200 US marines who were stationed in Nagasaki after the war was over and the occupation of Japan had begun. Arriving in the city in September, more than a month after the bombing, the soldiers were assured by the Army that the radiation contamination of the city was 'below the hazardous limits' and were therefore ordered to help in the clean-up of the rubble left by the bomb blast. One of those Marines, James McDaniel, recalls that despite the fact that the weather was cold with frost on the ground, 'the metal felt warm in my hands'.[253] Another soldier, Harry Coppola, remembers that 'You could take bricks, and they'd turn to powder. I used to kick it around. You'd kick it, and it'd turn to dust.' Both men have now developed bone-marrow cancer, and they believe it was due to their exposure to the residual radiation of the bomb. Coppola in fact got so sick at the time that because of his constant nausea and vomiting, the marine doctor finally sent him back to the US. McDaniel's and Coppola's experience is not unique. Writer Norman Solomon has researched all of the 1,000 Marines stationed within a mile of the blast area at the end of September 1945, and the incidences of bone-marrow cancer are far above the national norm.

It is estimated that the US military exposed another 200,000 to 500,000 American soldiers to different levels of radiation exposure during the testings of the nuclear devices in order to learn how they would perform under atomic war combat situations. During one blast in 1957, 2235 servicemen were exposed. Studies on as many of them as have been located by the National Centre for Disease Control indicate that their leukemia death rate is running at twice the national average. Another test during that same year exposed Marines at 400 metres from Ground Zero. A study of one particular soldier from the unit, Russell Jack Dann, shows that within a year he had lost his teeth and hair, his joints ached and his sperm count dropped. He is now a paraplegic. Studies are in progress on the rest of Dann's unit.

Perhaps the most complete study of the military use of nuclear weapons comes from what has been discovered among the civilian employees of the Navy at the Portsmouth, New Hampshire, Naval Shipyard. What is important about this incidence is that it is not involving either fallout or deliberate exposure. Rather, this concerns workers who were merely refuelling and doing routine maintenance for the atomic submarines which came in periodically to the shipyards. The study was conducted by Dr Thomas Najarian, a hermatologist at the Boston Veterans Hospital. He first became interested in the shipyard workers after a patient of his mentioned that many of his fellow workers were dying at relatively early ages of blood-related diseases. Najarian began to investigate this assertion, only to find that the Navy refused to release its medical records. So Najarian utilized the death certificates of all those workers who had died after having worked at the shipyard. He discovered that 38.1% had died of cancer – twice the national rate. He also discovered that the leukemia rate among workers who had been exposed to accidental spills of radiation was 450% higher than the national norm. Furthermore, 60% of those workers aged between 60 and 69 who had been exposed to radiation spills had died of cancer.

One of the workers, Ronald Belhumeur, who had been an employee for 30 years at the shipyard, told a Congressional subcommittee on 28 February 1978 about the aftermath of one particular spill from the atomic submarine, Nautilus:

> The crew I worked with on the Nautilus are all dead. I am the last one. The machinist died of cancer only months after the spill. My fellow workers, tank cleaners, died from cancer and one from natural causes. Then in 1977, my supervisor and a machinist supervisor died only months apart with leukemia. After the spill on the four of us and the supervisors went to see what caused the spill, they must have really got belted. When I say this, Mr Chairman, it is (belted) with radiation. . . . In my opinion it was rotten with radiation exposure.[254]

K. THE RATIONALE FOR NUCLEAR WEAPONS: DETERRENCE

The sacrifices made by the public both in terms of money and lives for nuclear arsenals that can annihilate humanity is all done in order to keep the balance of terror so great that the terror itself acts as a deterrence against war. As Henry Kissinger explains it:

> With no advantage to be gained by striking first and no disadvantage to be suffered by striking second, there will be no motive for surprise or pre-emptive attack. Mutual invulnerability means mutual deterrence. It is the most stable position from the point of view of preventing all-out war.[255]

Put more simply, this means that deterrence is the attempt to control the tensions between the superpowers by the threat of mutual assured destruction of both countries should either side commit a forbidden act. To work there must be an equal balance of terror on each side, a parity, so that each feels relatively secure. It is within this parity that a policy of détente can be worked out. Ultimately, deterrence based on the mutual assured destruction of both countries is predicated on both sides using essentially the same methods of rational analysis in deciding what parity is and what the forbidden acts are.

What needs to be pointed out, however, is that the very dynamic of the nuclear arms spiral is antithetical to the notion of a meaningful deterrence. Since the beginning of the nuclear age the dominant psychological characteristic has been the combination of paranoia and an identification with the atomic bomb; this due to a psychic numbing at the spectre of the death and destruction nuclear war really means. It has not been rational; it has not been content to maintain a parity. Rather, at each stage there has been the compulsion for the 'and something other' described by Molotov in 1949. An advance on the one side breeds suspicion and fear on the other as each side seeks desperately to circumvent the other's defences while perfecting its own. Since each side is making the same effort, both sides fear the 'enemy' will make the 'breakthrough' first. The more feverish the pace, the greater the mistrust and fear.

Furthermore, the entire thrust of this deterrence is towards making nuclear war 'credible'; it is not towards making an equal balance of terror. The drive towards credibility forces the spiral into such things as mini-nukes and the development of guidance systems so accurate that the missiles can be reasonably expected to come within ten metres of their targets. It is towards further refinements of what terror means and further developments in all the conceivable ways in which one can trick the enemy or be tricked, both sides knowing that the game they play is one in which a sneak attack can happen in seconds and minutes rather than hours or days. Deterrence is based, therefore, not upon either mutually assured destruction or upon a realistic parity; it is based upon the psychology of fear and upon the unsureness on both sides that they may not be able to launch a first strike without suffering unacceptable devastation in return. This is a fact even admitted by Kissinger, although in a statement written before his rise to power in the Nixon administration: 'Deterrence', he said, 'depends above all on psychological criteria. . . . For purposes of deterrence a bluff taken seriously is more useful than a serious threat interpreted as a bluff.'[256]

A concrete example of this observation can be seen in the present situation in Europe. According to Herbert York, Director of Defence Research under Presidents Eisenhower and Kennedy, NATO's plans for the defence of Europe rest neither on the notion of mutual assured destruction nor on that of parity. They rest upon an 'awesome bluff':

> On the one hand, NATO says that if the Soviets launch a massive attack on Europe, even without using nuclear weapons, then NATO will reply with nuclear weapons and thus set in motion a series of events which could easily result in the destruction of most of Europe's cities and the death of most of its people.
>
> On the other hand, most European political leaders, and probably most of its military leaders, would not in fact be willing to deliberately set in motion events leading to such a result. However, the public admission of such an unwillingness would, as the experts say, 'undermine the credibility of the deterrent' and thus increase the probability of an attack.[257]

Deterrence is inherently unstable, therefore, because it requires military muscle-flexing, real or feigned aggressiveness, and a reliance upon force rather than conciliation to keep the balance of terror in operation. It is unreliable, furthermore, because it is predicated upon the 'enemy' utilizing the same rational calculations 'we' have gone through in deciding what the forbidden acts are and what risks are worth it and which ones aren't, all within the psychological climate of paranoia completely antithetical to such rational calculation. The Cuban missile crisis in October 1962 is a classical case in which the policy of deterrence came within a hair's breath of degenerating into an intercontinental ballistic war between the US and the Russians, all over one single Soviet ship crossing an arbitray line the Americans had drawn around Cuba. The Soviets miscalculated what the Americans considered a forbidden act to be and therefore pushed the paranoia to the point of presenting an intolerable threat to the Americans. The American threat in return was misinterpreted by the Russians intially as a bluff.

Both caught on to the gravity of the situation barely in time. It was clearly a case in which both Kennedy and Khrushchev came eyeball to eyeball, both miscalculating and threatening to pull the trigger – and then Khrushchev blinked. Had he not, we could well not be here.

L. NUCLEAR PROLIFERATION

The nuclear arms spiral between the superpowers and their allies has had consequences far beyond the boundaries of the United States, Europe and the Soviet Union. It has been a spiral that has caught the entire planet in its spin, a nuclear arms race that has been accompanied by an equally spiralling *conventional* arms race. Both the US and the USSR have sought to arm other countries to the same degree they continue to arm themselves, although because they both wish to keep a monopoly on their nuclear arsenals, their arms sales are thus far in the non-nuclear weapons categories.

Nevertheless, this is a very lucrative business. According to Michael Klare, a foreign policy analyst at Princeton University's Centre of International Studies, Pentagon officials are literally told that the job of increasing arms sales abroad should be viewed as a 'mission'. Accompanied by corporate executives of companies doing business with them, therefore, Pentagon officials 'stalk the globe in search of prospective customers'.[258]

By all accounts, this campaign has been overwhelmingly successful. Between 1968 and 1975, arms exports under the foreign military sales programme in the United States rose over 1,200% – from $798,000,000 to $9,500,000,000, and the pace has continued at even a faster climb since then, making America the largest single arms supplier in the world. Between 1965 and 1974, the US delivered a total of $31,600,000,000 worth of arms to foreign countries, or more than 50% of all the arms traded on the international market. The US export of arms to the Third World in particular during this period amounted to $24,500,000,000, or 53% of all arms acquired by the Third World. In the last several years the conventional arms spiral has reached such an intensity that orders for new arms are in fact exceeding actual deliveries by more than three to one.[259]

By 1977, armament of all kinds on a world wide level had reached a total of $360,000,000,000. NATO and Warsaw Pact countries spent some 70% of this amount, the Third World

18%. In 1978, world-wide expenditures totalled over $400,000,000,000, and of the arms being shipped abroad rather than kept at home, 75% were going to the Third World.[260]

The rationale behind conventional arms sales is similar to that behind the continuing nuclear arms competition: it is to give the recipient countries a meaningful deterrent against attack. In situations where a power imbalance exists, it is argued, new arms deliveries can reduce the risk of war by eliminating any perceived weakness a country might feel.

Edward Luck of the United Nations Association, however, argues essentially what was stated in the previous section concerning the flaws in believing that nuclear weapons offer a 'credible deterrent'. He states that 'the massive transfer of armaments . . . can exacerbate local tensions and increase the likelihood of conflict', *not* stabilize the situation. Citing the 'explosive Third World areas' such as the Middle East, he observes that the introduction of 'innovative and highly sophisticated weapons . . . may magnify uncertainties in the perceived military situation, leading to overconfidence or insecurity between rivals'.[261]

Michael Klare illustrates this point by giving the example of the Arab-Israeli conflict. Suppose, he says, that Israeli intelligence sources report 100 new tanks have been delivered to Egypt. Under worst case planning, the Israeli defence staff will assume that 100% of the Egyptian tanks will be operational in battle but that the Israelis will be minus its usual 15% due to maintenance and repair. The net result is that Israel believes it needs at least 130 more tanks to redress the imbalance that now exists. But since a shipment of this many tanks to Israel would be in turn interpreted by the Egyptians as a new imbalance from their perspective, they would feel, based on calculations similar to that of the Israelis, that they would need at least 65 additional tanks to create a new 'balance'. 'Theoretically', says Klare, 'this process would continue until a significant imbalance occurred, and one side or the other felt obliged to launch a pre-emptive strike. It is obvious, therefore, that any effort to "balance" military capabilities in a high-tension area will automatically trigger an upwardly spiralling arms race while increasing the risk of conflict'.[262]

Without any restraints on the quality of the conventional arms being shipped abroad, this spiral can only result in the request and delivery of increasingly potent weapons to troubled areas, thus making each new conflict more devastating than the one before. This can perhaps be seen most clearly in the successive levels of destruction and battle fatalities in the four major Arab-Israeli wars.

This conventional arms spiral leads directly to the participating countries seeking nuclear weapons at some point. The conventional arms race has upper limits, and in terms of the increasing demand for more potent weapons, nuclear potency begins where conventional potency ends. Referring back again to the Arab-Israeli conflict, this natural evolution from conventional armaments to the desire for nuclear weapons can be seen. The situation has been summed up in an article appearing in *The Jerusalem Post* on 25 April 1976, and although it only represents the opinion of its author, Ephraim Kishon, it contains a logic that is compelling:

> From time to time the US Administration wonders . . . why we're so greedy. From time to time it plays dumb and pretends not to know of this tragic situation where three million weary Jews who've just begun building their home in the desert are being forced to maintain a huge military force to defend themselves against a hundred-million millionaires building up an army of NATO size. The US administration acts as if it had no idea that nearly half our Gross National Product lies under wraps in our military emergency stores, and that if it weren't for this back-breaking burden we wouldn't be standing like beggars at their door.
>
> All this generous American assistance, even when it's called economic, goes directly or indirectly to sustain a losing arms race. All the parties involved have an interest in this race, each for his own reasons – except Israel who can never win it. To be sure, Israel won't be defeated in battles: it'll collapse – economically and socially – under the fearful load of endless arms purchases. . . .
>
> It's a fully planned vicious circle: when the Arabs have 10,000 tanks, we'll need at least 6,000; when they have 20,000, we'll need 12,000 – and so on *ad infinitum*. Interim agreements or not, the race will go on, and our total dependence on the US. . . .
>
> Our one and only alternative to our gradual destruction by arms race is to develop a nuclear deterrent of our own. It's our single chance for telling our many enemies and our one friend: that's it, we're not playing any more, we refuse to go

> on running for ever in the circles you've drawn for us. We want no more of your arms, we want a sophisticated educational system.
>
> Sooner or later we'll have to say it out loud. Sooner or later we'll have to announce: if any Arab army crosses this green line we reserve the right to use atomic weapons, and if it crosses the red line we'll drop the bomb automatically, even if this whole country is blown up by nuclear retaliation. You don't believe it? Try us! . . .[263]

What is frightening about Kishon's article is that Israel does in fact have a nuclear arsenal even though officially denying it. What is even more disturbing is the way in which Israel obtained these weapons.[264]

As best it can be discovered, the Israeli decision to develop a nuclear arsenal came shortly after the 6 June war in 1967. Defence Minister Moshe Dayan, who put the proposal before the Israeli cabinet, felt that while Israel had indeed won a resounding victory this time, the point was sure to come sooner or later when the Arabs would arm themselves sufficiently seriously to threaten Israel's national existence. Although Prime Minister Golda Meir was at first resistant to the idea, feeling a natural repugnance against Israel possessing weapons of mass destruction, she at last consented, and the cabinet voted unanimously to go nuclear.

As all legitimate channels for obtaining either bomb grade uranium or plutonium were closed, Israel could only become nuclearized by covertly stealing what was necessary. This task was given to Israel's central bureau of intelligence and security: Mossad.

Mossad set up a special commando unit in either late 1967 or early 1968 whose specific mission was to raid the Western nuclear powers. Three European nations were targeted as well as the United States.

The raids occurred during 1968. In France, a twenty-five ton truck carrying government uranium was hijacked after firing tear gas canisters into the truck's cab to disable the driver. The radioactive booty was then smuggled over 2,000 miles to the Negev desert where Israel's Dimona nuclear installation is situated. The same operation was repeated in England, although the hijacked uranium turned out to be low grade rather than the required enriched uranium that the French hijacking had produced.[265] The most spectacular aspect of the affair was the hijacking of a freighter, the Scheersberg, with a cargo of 200 tons of uranium while it was on the high seas off the coast of Italy in November 1968. The boat was sailed to the port of Haifa, the cargo unloaded, and then sailed back again to its final destination in Europe – empty.[266]

Mossad's technique for stealing uranium in the US was somewhat more subtle. It was also an operation that had been going on for some time. In collaboration with officials of the Nuclear Materials and Equipment Corporation (NUMEC), most probably including its president Zalmon Shapiro, who through the Corporation had worked as a sales agent for the Israeli Ministry of Defence, the Israelis were able to siphon off an estimated 200 to 400 pounds of enriched uranium from NUMEC's Apollo, Pennsylvania, nuclear plant during the years 1957 to 1967.[267] When one considers that only about 25 pounds of enriched uranium are needed to make a functional nuclear bomb, the theft of between 200 to 400 pounds from the Apollo plant, in conjunction with all that was stolen from Europe, was enough to give Israel a rather sizeable nuclear weapons arsenal.

The American Central Intelligence Agency (CIA) became aware of Israel's nuclear weapons capacity in 1968. According to the former CIA deputy director of Science and Technology, Carl Ducket, the 'CIA had drafted a national intelligence estimate on Israel's nuclear capability in 1968. In it was the conclusion that the Israelis had nuclear weapons.'[268] According to Ducket, the CIA immediately informed President Johnson. His response, however, was to instruct the CIA not to tell anyone – not even the Secretaries of State or Defence.

By 1973, Israel's nuclear plant at Dimona had produced a small arsenal of nuclear bombs the CIA estimated to be in the 20 kiloton range, meaning bombs about 30% larger than the one dropped on Hiroshima. During the October War that same year, Israel came very close to using them.

This fourth Arab-Israeli war began with a surprise attack by both the Syrians and the Egyptians at 12:05 p.m. on 6 October 1973, as the Israelis were beginning their worship for the Day of Atonement, the holiest of all Jewish festivals. Caught totally unprepared, what Israel had always feared began to happen: she seemed to be suffering defeat at the hands of the Arabs. Indeed, the Arab armies did make strong early advances. They had planned the entire operation, condenamed Badr, very carefully, so carefully that Israeli intelligence remained unawares. The attack came on the tenth day of Ramadan, the day on which in AD 624, the prophet Mohammed began preparations for the Battle of Badr, his first victory in the long campaign that was to finally conquer Mecca.

In the first five days of fighting, Israel managed to neutralize Syria in the north but was unable to neutralize the Egyptians, who had managed to secure a massive bridgehead across the Suez Canal and occupy large areas of the Sinai desert. Israel was in deep trouble: her losses of aircraft and tanks had been enormous, as had battle fatalities. Moreover, ammunition was running out. In the tank corps alone there was less than four days stock of ammunition left. Reluctantly, therefore, Israel agreed to accept a US proposal for a ceasefire on 11 October – without even insisting that Egypt withdraw from the areas it was occupying in Sinai. Sadat refused the ceasefire proposal, however, thus placing Israel in the position of feeling that a nuclear attack upon the Arabs was the only option left.

At this point President Nixon began a massive airlift of supplies to Israel, not only because of pressure from the American public, but because Secretary of State Kissinger informed him that Israel, facing defeat, was at the point of resorting to nuclear weapons. This meant specifically that Israel had already equipped several surface-to-surface missiles with nuclear warheads. The missiles, two-stage solid fuel rockets designated MD-660s, had been obtained from France and were capable of hitting targets 280 miles away – sufficient therefore to hit Cairo, Amman, and Damascus.[269]

Fortunately, the airlift of supplies was enough for Israel to be able to snatch victory from the jaws of defeat, although it was a victory so costly that even in victory there was the smell of defeat. It was this way, not only because of the appalling cost in human lives lost (the Israelis lost 2,500 soldiers, the Arabs 16,000), but because Israel had been forced to resort to the nuclear weapons option, a move that if carried out was recognized by the Isrealis as holding forth the possibility of double suicide. The question also remains concerning how much President Nixon's massive airlift was prompted by a genuine desire to keep the Isrealis from defeat and how much was because of a perceived threat of nuclear blackmail, that if aid was not immediately forthcoming, Israel would use the bomb.

The case of Israel illustrates the point being made that not only is the conventional arms spiral as inherently unstable as the nuclear arms spiral but that because it has an escalating momentum towards more and more potent weapons, the conventional arms spiral ultimately leads to the desire for nuclear armaments.

The proliferation of nuclear arms comes from another source besides the conventional arms spiral, however, which brings into the issue a dimension not previously raised: *nuclear power plants*.

The connection between nuclear weapons and nuclear reactors is perhaps the most important aspect of the nuclear weapons proliferation problem confronting the world today, for wherever there are nuclear power plants operating – and nearly thirty countries have them as of 1979 – there is both the material and the technology to construct nuclear bomb devices.

This problem began in 1953, when President Eisenhower launched an 'Atoms for Peace' programme, specifically designed to share with the rest of the world the 'peaceful uses' of atomic energy. According to Paul Leventhal, an expert on nuclear weapons proliferation, 'the whole thing was at first conceived primarily as a public relations operation in the cold war. After the shock caused by the explosion of the first Soviet hydrogen bomb, Ike wanted to appear before the United Nations and make it plain that the Americans . . . were ready to put their knowledge, their skill, and their financial resources also, at the disposal of those who wanted to use nuclear energy for peaceful purposes'.[270]

Eisenhower made this Atoms for Peace proposal to the UN on 8 December 1953. In early 1954, the Congress passed an amendment to the Atomic Energy Act of 1946, authorizing the US to make the 'peaceful uses' of atomic energy as available as possible, subject, of course, to defence and security considerations. Within a year, 25 bi-lateral agreements had been signed contracting the US to provide countries with either nuclear power reactors or nuclear materials for research. In 1964, another amendment was attached to the 1946 Atomic Energy Act authorizing the AEC to license private American nuclear companies to own and export nuclear material.

At the same time as Eisenhower was making his Atoms for Peace speech in the UN, however, a memorandum was circulating inside the government, written by Robert Oppenheimer and Major General Leslie Groves, asserting that the impure plutonium which nuclear reactors produce was sufficient for the construction of simple but highly effective nuclear bombs.[271] This observation was rejected by the Eisenhower administration, Secretary of State Dulles insisting that nuclear bombs could only be made through 'complicated, difficult and expensive measures that could not remain undetected'.[272]

On 18 May 1974, India proved him wrong. In a top secret operation, codenamed 'Buddha', India successfully tested a nuclear device constructed by using fissionable material from an experimental reactor obtained from Canada, although using American heavy water. Using Atoms for Peace, India joined the élite group of nations possessing atoms for war.

That India was easily able to do this can be seen from examining the structure of the organization that is specifically mandated to monitor such activities. This is the International Atomic Energy Agency (IAEA), set up in Vienna in 1957 on the initiative of the US. It was given the power to inspect and inventory nuclear materials in order to allow it to detect any diversion of nuclear materials to military purposes. But the IAEA has no power to enforce its findings nor does it have the power to enforce its regulations. Countries participate in the IAEA voluntarily and can expel the IAEA from their borders if they believe it is attempting to perform surveillance upon them.[273]

With over forty-three nations possessing nuclear programmes in some stage of development, it is not surprising that several are already making moves to follow Israel and India's example. This is particularly true of Brazil and Argentina, both competing for mini-superpower status on the South American continent. Their competition to develop a nuclear device as part of this competition was specifically triggered by the Indian detonation. According to Norman Gall, writing in *Foreign Policy* 'each has regarded the other's activities in the field of nuclear energy with mistrust for some time. After May 1974, speculation about which of the two countries would have the bomb first became an ordinary subject of dinnertable conversation among the élites of both.'[274]

Brazilians such as Chancellor Gibson Barboza and diplomat Sergio Correada Costa have stated publicly that Brazil needs to develop nuclear weapons. It is not surprising, therefore, that Brazil has refused to sign an international agreement to keep Latin America a nuclear-free zone. 'For Brazil the bomb is a kind of military imperative,' says Brazilian journalist Murilo Mello Filho, 'a political necessity. . . . The current world situation allows no lighter or more agreeable choice.'[275]

It is thought by some experts that Argentina is closer to exploding its first nuclear device than Brazil is. It submits to inspection by the IAEA but has not let these inspections slow its weapons programme. Indeed, according to a *Newsweek* article of 7 July 1975, 'US intelligence sources reported last week that, without IAEA detection, Argentina recently removed fifty kilograms of plutonium waste from its Atucha station – potentially enough for five atomic bombs.' At the time Argentina was attempting feverishly to acquire its own re-processing plant. Like its neighbour to the north, Argentina considers the Bomb a political necessity. One Argentinian journalist, Mariana Grondona, writing in *La Opinion* in December 1974, expressed the point this way: 'Now with India the atomic powers are six in number. They would like to remain six. If we come to be seventh, we would of course like to see no more than seven. The last one to arrive tries to close the door. That is only natural. But why should the door be closed in our face?'

Similar logic was used in Taiwan following the Indian explosion. The 27 February 1977 edition of *The Washington Post* stated that, 'Behind barbed wire and guarded gates teams of Western-trained scientists are quietly but relentlessly nudging Taiwan into the nuclear arena. . . . Ensconced in the Taiwan institute is a forty megawatt Canadian-supplied research reactor closely resembling the reactor that was instrumental in India's 1974 development of its first nuclear device.' Taiwanese Premier Chiang Ching-kuo has even admitted publicly that Taiwan has the capability of developing nuclear arms. The Chung Shan Institute has full scale programmes for missile design, rocket fuels, nuclear warhead research and projectile guidance systems.

Perhaps of most concern recently have been the growing indications that Pakistan is about ready to explode a nuclear device. In the summer of 1979 it was confirmed that Pakistan had built a uranium enrichment facility at Kahuta, not far from the capital city of Islamabad.

There are two reasons for particular concern about Pakistan's nuclear development. The most important one is that Pakistan is considered to be one of the most underdeveloped countries in the world and yet is still proving itself capable of developing a nuclear capability entirely on its own, without the help of an industrialized nation. This has grave implications, for if Pakistan can do it there is nothing to stop virtually any other country with the political will from doing the same. The second cause for concern centres on the fact that the bomb Pakistan develops may be made available to other countries, particularly Islamic nations, perhaps even the Palestine Liberation Organization. The late President of Pakistan, Zulfikar Ali Bhutto, who began Pakistan's nuclear programme, referred to it as 'the Islamic bomb'.

Bhutto made this decision to go nuclear in 1973, when it became clear that India was in the process of developing its nuclear bomb programme. His rationale was threefold: first, that Pakistan had no real intention of manufacturing bombs themselves, being interested instead in developing nuclear explosives for 'peaceful' purposes such as digging canals, removing mountains, etc.; secondly, that even if Pakistan in fact built a nuclear bomb, neither the US nor the Soviet Union had grounds for complaint since in the 1968 Non-Proliferation Treaty they had agreed to negotiate 'seriously' for disarmament and had not; and thirdly, that Pakistan was entitled to nuclear bombs because India had them.

India had offered essentially this same rationale in 1975 when it exploded its nuclear device, K. Subrahmanyan, Director of the Institute of Defence Studies in New Delhi, stating: 'You and I are asked to accept the credibility of the structure of peace built on 7,000 strategic nuclear warheads; in addition to another 7,000 tactical nuclear weapons capable of incinerating all of us on this globe many times over; the credibility of a Non-Proliferation Treaty since the signing of which nuclear weapons have quadrupled in number; the credibility of deterrence which means a non-stop arms race.' The argument, when put simply, is this: if the US and the Soviet Union can carry on the nuclear arms race unchecked, why cannot the rest of the world?

Indeed, in October 1979, there was satellite detection of an apparent atomic flash in the southern Indian Ocean with subsequent identification of radioactive fallout in New Zealand. The location of the flash off the southern coast of Africa fuelled speculation that the explosion was an attempted covert experiment in South Africa's nuclear programme, although some American authorities have argued that because of nuclear weapons manufacturing and delivery capabilities that are known to be more advanced, and because of more compelling reasons for secrecy, the bomb was most probably Israeli.

Plutonium, essential to the construction of nuclear bombs, is already being produced by reprocessing facilities in Mühleberg, Switzerland; Biblis, West Germany; Latina, Italy; Vandellos, Spain; Barseback, Sweden; and Atucha, Argentina.

According to Albert Wohlstetter, a leading authority on the subject of proliferation, there are as of 1979 at least eighteen countries besides the six officially nuclear nations – the US, Russia, Britain, France, China and India – which have enough plutonium for several bombs. He expects that by the year 1985, forty countries will be in a position to manufacture atomic bombs, given the political will to do so.[276] This point needs to be stressed: *every country with a nuclear programme is capable, if the political determination is there, to use its Atoms for Peace in the construction of atoms for war*, especially if it has a reprocessing plant and a particular tendency towards aggressive foreign policy. What is alarming in this respect is what proliferation expert Thomas Cochran referred to in testimony at the 1977 Windscale Inquiry as a 'binary operation'. This involves countries, believed by experts to include South Africa and Israel, which keep all the component parts ready but not brought together. This means that they are technically non-nuclear but capable of becoming so in days or even hours if the situation demands it. This can lead to conditions such as were reached during the October War between Israel and the Arabs in which one side, not knowing that the other is in fact in possession of nuclear weapons, will do something that forces what began as a conventional confrontation into a nuclear holocaust or the possibility thereof.

So widespread are both the Atoms for Peace programme and the political determination to connect this programme with a nuclear weapons programme that Robert Jungk estimates that there will be enough nuclear explosives scattered around the planet by the year 2000 for 1,100,000 atomic bombs of various sizes, if the atomic programmes of the various countries are continued. This estimation gives weight to the observation of David Rosenbaum, one of the first people to write on the dangers of the malevolent use of 'special nuclear materials', that 'the sad truth is we've opened a Pandora's Box and there is no way to return to a world safe from nuclear weapons, even nuclear weapons in the hands of terrorists. Atoms for Peace may turn out to be one of the stupidest ideas of our time.'[277]

This 'sad truth' becomes tragic when it is recalled that the rapidly escalating conventional and nuclear arms spirals are taking place in the midst of crushing human poverty. In 1975, while the world collectively spent *$400,000,000,000* arming itself, *ten million people* starved to death. In 1976, the amount the world again spent arming itself was equal to what it spent on health and more than it spent on education. This was in spite of the fact that 570,000,000 people remain undernourished, about 2,800,000 people are without safe water – water-borne diseases kill an estimated 25,000 persons every day – about 1,000,000,000 people lack adequate housing, about 1,500,000 are without effective medical care, about 250,000,000 children do not attend school and 800,000,000 people are illiterate.[278] In June 1980, the US Pentagon announced that it will spend $1,000,000,000,000 (one *trillion* dollars) over the next five years on 'defence'. During the same month, the US Presidential Commission on hunger announced that 50,000 people die of either starvation or hunger-related diseases *every 24 hours* somewhere around the world.

The gravity of the situation can be realized another way. In 1973, the world military expenditures came to $245,000,000,000. This was 163 times more than the amount spent on international co-operation and development through the United Nations system. The world's military expenditures are today greater than the gross national product of all Africa and South Asia combined. The enormity of this can be seen in the fact that 50,000,000 people are employed for military purposes around the world. Furthermore, close to 50% of all the world's scientists are in some way involved in military research and development, a scientific endeavour that costs between $25,000,000,000 and $40,000,000,000 per annum. These sums represent 40% of all the public and private research going on anywhere on the planet.[279]

These short-term statistics, however, must be put in the context of a longer-term crisis between the rich and the poor. It is true that between 1965 and 1975 the percentage of Third World expenditures for arms rose from 6% of the total to 17% and the percentages are still rising. Yet this rise is largely because the Western countries, being forced through political revolution to give up the imperialism of colonies, have replaced political tutelage with economic tutelage, creating markets in the Third World for commodities the Western industrial economies wish to sell. Recall Michael Klare's remark that the Pentagon sees its job of developing new markets for its new weapons as a 'mission'. The intensity of the commitment to this 'mission' can be seen in the enormous bribes companies with heavy Pentagon connections such as Lockheed are willing to make to ensure that their weapons are bought instead of those of their competitors. While the official legitimation is that of selling weapons to ensure peace through credible deterrents, the underlying reasons are to secure a favourable balance of trade for the exporting country; to ensure full production and thus full employment in the industries and for the corporations that have their mainstay in defence contracts for new weapons; and to ensure that because the production runs of these weapons systems are continued and expanded, the exporting country pays even less for its own procurements. So vital is the export of weapons for profit, particularly to the burgeoning Third World markets, that Deputy Secretary of Defence William Clements has asserted that any restriction on weapons exports 'decreases the potential contribution of sales . . . to strengthening both free world security and the US economy and balance-of-payments position'.[280]

Because the world's poorest countries are caught up in the economic markets of the military-industrial complexes of the Western nations, they are forced into a situation in which they must neglect the basic task of addressing the needs of their starving millions. And therefore the poor become poorer and the rich richer. In 1970, for instance, this can be seen: the world's richest 1,000,000,000 people earned an income of $3,000 per person; the world's poorest 1,000,000,000 people earned not even $100 each. By 1980, the $3,000 of the rich is expected to climb to $4,000; the $100 for each of the poor is only expected to rise to $103.[281]

What this means is that well over 75% of all the world's income, 75% of all the world's investment and services, and almost all of the world's research and development are concentrated into the hands of less than 25% of the world's population. This 25% consumes 78% of all the world's minerals, and for armaments alone, as much as the rest of the world combined. This 25%, moreover, have constructed an economic system based on a permanent war footing, what Seymour Melvin calls a 'permanent war economy'. It is upon this basis that their wealth is made, that their power is continued, and current world 'peace' is established. With an ever increasing number of nations joining the nuclear club, however, and with the global tensions mounting each day, the probabilities of this peace based on terror erupting into nuclear conflagration increases proportionally to the intensity of the paranoia and anxiety felt and the number of nuclear bombs armed and ready for action.

M. THINKING THE UNTHINKABLE

It has been pointed out that the design, construction and ongoing expansion of the nuclear weapons programmes among those nations involved is based upon one primary concern: national security. The United States and its NATO allies have constructed a nuclear shield of over 30,000 tactical and strategic weapons for this purpose. Behind this capability to inflict planetary annihilation should they be sufficiently threatened, these industrial democracies feel that they are somehow 'safe', although never safe enough to stop building more weapons. The same posture of security through the ability to inflict totally unacceptable damage to the 'enemy' holds true for the Soviet Union and its allies as well.

But is this peace based on terror really an adequate insurance for peace? Dare we think the unthinkable, that nuclear weapons are *not* in fact the ultimate security blanket behind which the world can rest assured that the holocaust will never come? Dare we think that they will sooner or later bring down the very catastrophe that their presence is supposed to be insurance against?

When confronted with the spectre of annihilation, people tend to close themselves off and like the survivors of Hiroshima and Nagasaki retreat behind the barriers of psychic numbing – unwilling to recognize the truth and therefore incapable of stemming the problem.

In 1960, Herman Kahn wrote his classic work, *On Thermonuclear War*, which attempted, he said, 'to direct attention to the possibility of a thermonuclear war, to ways of reducing the likelihood of such a war, and to methods for coping with the consequences should war occur despite our efforts to avoid it'.[282] Two years later, Kahn wrote *Thinking the Unthinkable*, in which he stated that much of the criticism he had received for writing his first book dealt not with the correctness of his data or the soundness of his arguments but concerned whether a book on nuclear war should have been written at all. 'It is characteristic of our time,' he said, 'that many intelligent and sincere people are willing to argue that it is immoral to think and even more immoral to write in detail about having to fight a thermonuclear war.'[283]

To illustrate the point of the psychic paralysis many fall into when confronted by the nuclear war potential of the superpowers, Kahn cites the problem of white slavery in Britain at the close of the nineteenth century. Each year thousands of young women and girls from all levels of British society were forced into bondage and shipped to the brothels of Europe. One reason why this practice lasted as long as it did, Kahn suggests, was that it simply could not be openly talked about in Victorian society; moreover, the extreme innocence considered appropriate for British women made them easy victims, helpless to cope with the situations in which they became entrapped. Victorian standards, therefore, not only perpetrated the white slavery by refusing to talk about it, but intensified the damage done to those who were victimized.

This is not an isolated example. In January 1939, the oddsmakers in London were laying 39 to 1 odds *against* war breaking out in Europe that year. Human beings, like hibakusha, have an overwhelming tendency towards a psychic closing-off to situations they would prefer not to face. And yet ignoring the problem generally only makes it worse until it explodes. So, too, with the Age of Overkill and the plutonium economy which connects atoms for peace with atoms for war: the longer the problem is not dealt with the worse it gets. In essence, therefore, we act like the ancient kings who punished messengers with bad news. The result was not that the news changed, only that its delivery was slowed up.

With tens of thousands of nuclear missiles poised to strike, with bomber fleets on twenty-four hour alert and always airborne with armed nuclear warheads, with schools of submarines patrolling the ocean depths in the same state of alertness, with constellations of satellites directing delivery, and with an ever-increasing number of nations joining the nuclear club through their atoms for peace programmes – all within the context of a spiralling conventional arms race and ever-increasing international tensions, thermonuclear war may strike one as immoral, unthinkable, insane, even intolerable but, as Max Lerner points out, 'the intolerable is not by that fact impossible'.[284]

One has only to look back a mere four decades to the Nazi Final Solution of the Jews to recall a time in history when the absolutely intolerable, even unimaginable, occurred. Remember Dachau? There 283,000 people – mostly Jews – were gassed and their bodies burned. And Auschwitz? There 4,000,000 people – again mostly Jews – were gassed and burned – all for the sake of 'purifying' the Aryan race.

The intolerable can happen. True, it is absurd, but reality can be stranger than fantasy, and in politics and the historical movements of peoples and their leaders, there seems to be an even greater element of the absurd than in the rest of life. What could be more absurd than the United States being able to destroy completely every Soviet city over 100,000 *forty-three times over*? What could be more absurd than the Soviets developing killer satellites so that they can blow up 'enemy' communication systems in outer space? And what could be more absurd than the selling of nuclear reactors all around the world when each reactor produces enough plutonium each year for several Hiroshima-sized nuclear bombs?

And yet this absurdity is not only occurring but is actively supported by citizens' tax dollars and a propaganda campaign that calls such absurdity patriotism. Indeed, so psychically numbed have we become that we have allowed human history to degenerate to the point where if the nations were to use their nuclear arsenals against each other in a Third World War there is a high probability that not a single human being would survive. The Pentagon calls such a scenario 'wargasm'.

Fortunately, the prospect of this occurring has thus far kept nuclear weapons unused. Our paranoia has been balanced by our guilt and instinct to survive. But there are no built-in controls to ensure that this moral revulsion and consequent restraint will keep our paranoia and agressiveness in check indefinitely. As mentioned earlier, the notion of mini-nukes and the concept of limited nuclear war, along with the growing number of nations developing weapons of mass destruction, mitigates against this 'peace' being ensured much longer.

Indeed, it is within the context of ever-growing human deprivation amidst growing armament expenditures, of clashes between economic and political systems, of social instability and diminishing energy resources, that one should think about the unthinkable. Moreover, as more and more countries convert their atoms for peace into atoms for war, many may not play by the rules established by the superpowers: not necessarily out of any sense of inhumanity – what Lerner calls the 'Hitler factor' –

but perhaps simply out of a conviction of the country's own manifest destiny or sense of historical rightness: Israel, for example, or any one of the Arab states or the Palestinian Liberation Organization. Or South Africa, or Brazil or Argentina, or Pakistan or Taiwan. Any of these countries could for their own particular reasons choose to produce and to use nuclear weapons, given the 'right' historical circumstances. The United States already found one such 'right' situation in Hiroshima and Nagasaki. What is to keep other nations from finding their own?

Lloyd Dumas, in an article entitled 'National Security in the Nuclear Age', makes the point that while nations are seeking national security from the possession of nuclear weapons, there are four factors that mitigate against these weapons giving the security the nations are seeking. They are: the risks ensuing from nuclear accidents; the threat of an accidental war; inventory control problems; and the ever-widening gap between offensive and defensive weapons.[285]

1. Major Nuclear Accidents

Although any attempt accurately to assess the likelihood of nuclear accidents or even to give an account of the ones that have already occured must be aware of the heavy veil of secrecy around the nuclear weapons programme, it is still possible to discuss the problem with reasonable certainty. Examples must come mostly from the NATO countries, however, as the veil of secrecy is even heavier in the Soviet camp. It is assumed, however, that what is said about one camp applies to both.

Milton Leitenberg, a nuclear weapons expert in the Stockholm International Peace Research Institute, has estimated that prior to 1968 alone there were 33 major US accidents involving the total destruction of a nuclear weapons delivery system and the destruction, loss and/or other involvement of the nuclear weapons aboard that delivery system.[286] Twelve additional accidents are attributed to British, French or Soviet sources.[287] This is dangerous for two reasons, says Leitenberg: first, because it could start a nuclear war; and second, because it could do tremendous damage wherever the accident occurred. Major accidents include the following:

(*a*) On 23 January 1961, a B-52 bomber carrying two 24-megaton nuclear bombs crashed near Goldsboro, North Carolina. According to Dr Ralph Lapp, former head of the US Office of Naval Research, one bomb was removed from the wreckage, the other from a field nearby where it had fallen without exploding. When the recovery team examined this second bomb, however, they discovered that five of the six safety interlocks had been triggered by the crash. Only one single switch had prevented the explosion of a 24-megaton nuclear bomb. Had it exploded, there would have been a fireball not a half-mile wide, as at Hiroshima, but *two miles* wide, destroying all standard housing within 12.5 miles and igniting all flammable materials within 34.5 miles [288]

(*b*) On 17 January 1961, another B-52 bomber, this time carrying four 20-25 megaton hydrogen bombs, crashed near Palomares, Spain. One landed undamaged, although the conventional detonating devices of two others exploded, scattering plutonium over a wide area. This necessitated the removal of 1,750 tons of radioactive soil and vegetation. The fourth bomb fell over the Mediterranean and was only recovered after an intensive three-month-long underwater search.[289]

(*c*) On 21 January 1968, another B-52 bomber with four-megaton class hydrogen bombs crashed into Thule Bay, Greenland. Some bomb fragments were found, along with evidence of low-level radiation, but neither the plane nor the four bombs were ever recovered: they had melted through the seven foot thick ice and sank into the bay.[290]

Dumas estimates that another three major accidents occurred between 1970 and 1973, of equal magnitude to the ones just cited. In addition, twenty accidents involving total destruction and six serious destruction of the delivery systems occurred. This figure does not include the more than 150 accidents involving total loss of 'lesser nuclear capable delivery vehicles' which occurred.[291] All told, Dumas estimates that between 1950 and 1973, there were at least 63 serious accidents to nuclear weapons and their delivery systems. Only four of these are attributed to either the Russians or the Chinese. If one assumes that behind their barrier of secrecy at least half as many major accidents occurred as has happened to the US and its NATO allies, this would put the total number of accidents at about 90, or an average of one major accident every three months for the twenty-four years between 1950 and 1973. The probability is slight that a bomb will ever detonate, but given the fact that most are considerably more powerful than the Hiroshima bomb, when and if one does indeed explode 'accidentally' the consequences will be terrifying.

2. Accidental Nuclear War

Accidental nuclear war can be defined as 'an exchange of weapons of mass destruction not initiated with the purposeful calculation of the governmental decision-makers in authority'.[292] There are two focal points for analysis here: first, the nature and probability of triggering events; and second, the circumstances within which such events are likely to lead to war.

The most plausible and probable triggering event would be faulty communications systems. The most important background circumstance would be a context of international or bi-national tension in which the nuclear forces are being operated on a 'hair-trigger' basis. This was certainly the case during the Cuban missile crisis.

During periods of tension, an accidental nuclear war could be perhaps most easily triggered if an 'enemy' missile landed on 'home' territory, although accidents involving nuclear *capable* delivery systems might have the same effect as it is unreasonable to assume that a nation against which several nuclear capable delivery systems had been accidentally launched would wait until they had actually landed to respond, seeing first whether the delivery systems were in fact carrying nuclear warheads.

There have been at least five publicly known incidents in which US nuclear capable missiles overflew their target points and crashed near the territory of another country. Among these was the overflight of Cuba by a Mace missile in 1967 and the crash into the Straits of Taiwan after an aberrant flight towards China from the direction of Taiwan by another missile.[293]

There is also the possible failure of early warning systems. On 5 October 1960, for example, the central defence room of the North American Air Defence Command received a top priority message from the US Ballistic Early Warning System in Greenland, indicating that a missile attack had been launched against North America. The Canadian Air Marshal was immediately contacted to verify the message, and it was only after some fifteen minutes of intense waiting that the warning was shown to be false. Apparently the radars had echoed off the moon. Another example involves the US Submarine Emergency Communications transmitter buoys which on at least two occasions in 1971 were accidentally released from US Polaris nuclear submarines and signalled that the submarines had been destroyed by enemy action.

Whether an accident such as those just described would be enough to trigger a retaliation, limited or otherwise, is not clear. What is clear is that the nuclear forces on both sides as well as the accompanying communications and warning systems are specifically designed for quick response. Missiles launched in Moscow would hit American air space within less than half an hour and would be on target within ten to fifteen minutes after that. Time, therefore, and quick decision-making are at a premium.

Accidents within the context of hair-trigger response systems, particularly in times of crisis such as the Cuban missile crisis, or the tensions along the Sino-Soviet border in 1969, or during the worldwide US nuclear alert during the Arab-Israeli war in October 1973, would at least serve to escalate the tensions and mistrust, if not do actual damage. That they could lead to nuclear exchanges must also be kept as a real possibility, particularly given the current escalation in the numbers of countries possessing a nuclear capability.

3. *Inventory Control*

Inventory control involves two separate but interrelated functions: first, the protection and accounting of nuclear weapons in stock; and second, the detection of missing 'inventory items' should that occur.

The Nuclear Regulatory Commission defines nuclear materials unaccounted for (MUF) as:

> The algebraic difference between physical inventory and its concomitant book inventory after determining that all known removals (such as accidental losses, normal operational losses and authorized write-offs) have been reflected in the inventory.[294]

Considering the extreme dangerousness of nuclear materials, one would expect any materials unaccounted for to be kept at near-minimum levels. Such is not the case, however; the Nuclear Regulatory Commission (NRC) has had a MUF rate of at least 0.5% throughout its entire inventory history.[295] A study done by the AEC before the existence of the NRC indicated that the capability of its inventory control system to even be able to detect a 0.5% inventory loss in its plutonium stocks was less than 50%. It is hardly surprising, therefore, that in 1978 the NRC announced that over a *ton* of plutonium was 'missing' and over four tons of other radio-active materials were missing as well.

It is within this context that it can easily be believed that a nation such as Israel could siphon off some 200 to 400 pounds of enriched uranium from the Apollo nuclear plant over a period of ten years and remain undetected, particularly if it had the active collaboration of the plant's management.

If this 0.5% MUF rate is reduced by a factor of five, meaning to the assumption of a 99.99% complete knowledge and control of inventory, and applied to the nuclear weapons inventory, this would still leave fifteen warheads 'unaccounted for'. This includes only the 8,000 US strategic warheads and the 7,000 tactical warheads the US has in Europe, not the more than 15,000 other warheads that are scattered around the rest of the globe. This number does not account for the Russian, French, British, Chinese, Indian or Israeli arsenals either. This is not to say that the fifteen warheads unaccounted for are lying in an empty lot somewhere; rather, it is to say that their actual status and location are not known with any certainty. Since *one* weapon is enough to kill hundreds of thousands of people if detonated, to have fifteen unaccounted for is a serious failure in military preparedness and national security.

This failure to account for all weapons and/or all nuclear materials can become acutely serious in situations where the complications of blackmail and embezzlement arise. An example of this situation can be seen from what happened to the city of Orlando, Florida, on 27 October 1970. The mayor received a note demanding $1,000,000 and a safe escort out of the country in exchange for not blowing up the city with a hydrogen bomb. Ransom instructions as well as a diagram which experts confirmed as a workable model for a nuclear explosion were included in the note – as well as the statement that the nuclear materials used for the bomb had been stolen from AEC shipments. The AEC was immediately contacted, and because it could not give absolute assurance that the materials had *not* been stolen, the city was forced to pay the ransom. The blackmailer was caught finally and turned out to be a fourteen year old honours student with no apparent access to nuclear explosives, just the know-how of atomic bomb design from documents available publicly.[296]

In terms of actual weapons, the situation is such that, according to former Senator Stuart Symington, in 1970 US tactical nuclear weapons were not guarded properly 'at least in some places' and could therefore be seized.[297] In July 1974, during the height of the Cyprus crisis, the US became concerned for the security of the hundreds of tactical nuclear weapons it had both in Greece and Turkey. The President therefore ordered the US Sixth fleet 'to be prepared to send in a Marine detachment to recover the atomic warheads' should the necessity arise.[298] This situation would become quite serious, according to Admiral LaRocque, if 'we couldn't get those weapons out and we might have to fight our way in to get control'.[299]

Another example that causes concern in many quarters is the location of many of the warhead arsenals. For instance, in South Korea, US nuclear warheads are stored as close as thirty-five miles south of the Demilitarized Zone. What would ensue if they were overrun by the North Koreans? Even in Europe, General Michael Davison, Commander of the US forces in Europe, admitted in 1974 that his troops would have difficulty protecting several of the existing nuclear weapons arsenals against terrorist attack.[300] A Defence Department report in that same year revealed concern that inspections were failing to report deficiences in these arsenals; that nuclear weapons were being stored out in the open; that equipment was being left in restricted areas capable of carrying the weapons away; and that the alert aircraft areas were often inadequately lighted and fenced.[301]

When even the degree of control theoretically possible falls short, it is certain that the degree of control in practice will be totally inadequate – and the bombs being left thus unguarded are weapons of unparalleled mass destruction.

4. The Gap Between Offensive and Defensive Weapons

In nuclear jargon there are two kinds of defence: defensive defence, meaning the protection of the defended barriers by static barriers and/or the active destruction of incoming forces; and offensive defence, meaning a pre-emptive first strike which destroys the enemy's forces before they can attack either first or in retaliation.

The fundamental problem with defensive defence is that it is by nature reactive, leaving the critical choices of method, timing and location to the attacker. Thus the defensive posture must be always ready for anything and capable of responding within the time restrictions of the warning system.

As late as World War II, static defence systems could be constructured which could be only destroyed by persistent offensive bombardment. This made it possible for defenders to defeat an attacking force by only destroying a small percentage of their forces. During the Battle of Britain, for example, an interception rate of only 10% of the incoming Luftwaffe bombers was considered necessary for a British 'victory'. This rate meant that each bomber would only be able to carry out an average of ten strikes, not enough to do intolerable damage.[302] With ratios such as these, the defence had important advantages.

Not so with nuclear weapons. There is simply no way to create a workable static or active defence against their onslaught, the reason being that the destructive impact of each weapon is so enormous that for any defence to work it must be able to destroy not 10% of the attacking missiles but 100%. The requirement is that of perfection, and perfection is not achievable.

This understanding came to light with particular clarity during the 1960s debate about whether to adopt an anti-ballistic missile system. After extensive debate throughout the military and civilian sectors of the decision-making process, the conclusion was reached that such a system was in the final analysis unworkable. It was an understanding that was shared by the Soviet Union as well.

The result of the agreement not to develop an anti-ballistic missile system, however, served only to encourage the Soviets and the Americans to build up their *offensive* defensive capabilities – called 'counterforce' by the Pentagon planners. This means the ability pre-emptively to destroy enemy nuclear forces before they can be launched to counter-attack. With no workable defensive system to withstand attack, therefore, the nuclear arms competition has focussed on the ability to mount such a devastating *first strike* attack that the enemy has nothing left with which to retaliate. The requirement for the first-strike capability, like that of defensive defence, is that of 100% knockout, for if only a small fraction of the enemy forces are left intact, they will be able to inflict unacceptable damage to the attacker. That a first strike could ever come even close to this 100% knockout becomes as obviously remote as the defensive defence doing so when one becomes aware that both the US and the Soviets have a *triad* nuclear force, meaning nuclear bombers, a good percentage of which are always in the air, land-based missiles and submarines and surface ships in and on the ocean. All three aspects of the enemy triad would have to be completely wiped out to ensure the attacking country's ultimate suvival.

It has already been pointed out that the new generation of US land-based missiles, if allowed to be built, will be mobile along underground tracks, making them virtually impossible for the Soviets to attack with any precision. Assuming, however, that in a first-strike attack the Soviets managed to knock out 99% of all the land-based missiles in the United States, including the new mobile MX missiles. This would leave fifteen missiles still untouched. Just these fifteen, a mere 1% of the US intercontinental ballistic missile force, are still capable of completely destroying up to 150 Soviet cities and/or military targets.

The air force part of the triad is perhaps the most vulnerable to attack, both because it takes time for the bombers to get into the air and because the planes and the weapons must be stored in or near airfields, allowing the enemy to know fairly precisely where they are. The US alert system is such that within fifteen minutes most bombers could be in the air. Fifteen minutes is about equal to the maximum possible warning time for an enemy missile attack, so it is safe to assume that many of the bombers would never make it off the ground. As bombers capable of carrying nuclear weapons are stationed all around the globe, however, for the Russians to attempt completely to knock out this part of the American nuclear triad would require attacks on several nations simultaneously.

Nevertheless, assuming the Soviets could destroy all the missiles and all the bombers in the US nuclear deterrent force, this would still leave the third leg of the triad – the surface ships and the nuclear submarines. Every US navy ship capable of carrying nuclear weapons does so. Because they are easily observed, however, they, like the bomber fleets, are vulnerable. Attacking the submarines, however, is virtually impossible. According to Richard Garwin, former advisor to the President for Science and Technology, 'the simultaneous tracking of 30 to 50 modern submarines is so difficult that its feasibility seems doubtful'.[303] Herbert Scoville, former Director of Research for the CIA, affirms this: 'Today it is difficult, if not impossible, to destroy even a single nuclear missile submarine that follows skilled evasion tactics.'[304]

When one considers that each of the Poseidon submarines in the US fleet can deliver more destructive force than all the bombs dropped by all parties during all the years of World War

II, and that the newer submarines, the Tridents, are capable of delivering between 48 and 224 entire city-destroying warheads, even *one* submarine left after destroying *all* the other submarines, surface ships carrying nuclear weapons, bombers, and *all* the land based missiles *simultaneously*, would be enough to inflict completely unacceptable damage upon the attacker. Such is the sophistication and horror of the nuclear warfare of the 1980s.

Modern weapons of mass destruction and the delivery and communications systems they are interlocked with are enormously complex. In order to prevent accident, sabotage, theft or war, they must be operated perfectly in order to guarantee reliability; this because even one weapon's destruction is too devastating to be allowed. In such complexity requiring perfection, the malfunction of the smallest part can affect the whole. Between 1966 and 1968, for example, the US Air Force carried out a series of readiness alerts on its Minuteman missiles in North Dakota. All failed – one because of a sub-standard resistor in the launch power supply; one because of a failure of a single capacitor in the guidance and control system; and another because of a faulty pin in one of the umbilical connectors.[305]

When one considers that the American nuclear triad is scattered all around the globe, the geographical dispersion and concomitant transportation, communication and security systems that such dispersion implies creates the need for extraordinary complexity and sophistication, particularly when this American nuclear deterrent must be co-ordinated with those of the French and British nuclear systems. The greater the complexity, however, the greater the probability of malfunction.

Perhaps the greatest area of possible malfunction lies with the *human* element rather than that of the computer or the machines involved. Imagine being in a nuclear submarine for the three months it remains submerged beneath the ocean surface – never seeing the light of day, living in crowded claustrophobic conditions, and realizing that you are part of a crew that could be ordered to launch missiles of mass destruction. Or imagine the conditions involved in descending each day into the missile silos of the land-based aspect of the nuclear triad, particularly in the light of statements such as that of a young officer at one such missile base near Omaha, Nebraska. 'We have two tasks,' he said, 'The first is not to let people go off their rockers. That's the negative side. The positive one is to ensure that people act without moral compunction.'[306]

To ensure the success of these two tasks, each member of the missile crew is provided with a pistol and given instructions to shoot anyone who appears likely *either* to fire the missiles without authorization *or not* fire them if authorized.[307] The strain upon a person knowing that each of the members of the crew has a weapon he is ordered to use on anyone unable to push a button that will unleash nuclear war is enough to produce extraordinary mental and emotional stress. This makes it mandatory, therefore, that each person in any way involved with the triggering of nuclear weapons be given frequent psychiatric tests.

In 1972, approximately 120,000 people had access to either nuclear weapons or the nuclear weapons release process in the US alone. Some 3,647 of them were released because of a perceived imbalance in their overall disposition, due either to mental illness, alcoholism, drug abuse or discipline problems. From 1971 to 1973, 1,247 NATO personnel associated with nuclear weapons were removed from their posts for similar reasons.

Human reliability problems are clearly a threat to the avoidance of nuclear accidents, nuclear inventory control, accidental nuclear war. When one considers that these weapons are primarily offensive, this problem is magnified, for it means that any malfunction, whether by machine or human being, will in all probability be directed at the 'enemy', thus raising the risk of retaliation.

What has been discussed so far are different factors that may lead to either nuclear accidents or accidental nuclear exchanges or wars. What still needs to be discussed is another 'unthinkable' dimension of the nuclear arms spiral: *deliberate* nuclear war. This is what the military chiefs of staff of the Soviet, American, NATO and Warsaw Pact countries spend most of their time planning for; it is what the weapons themselves are deployed for; and it is what has the most probability of happening.

It may be useful to contemplate the results of a 'limited nuclear war' in Europe, using only medium tactical nuclear weapons of one megaton, as a way of sensitizing oneself not only to what 'limited' means in nuclear terminology but what can occur when people are trained to operate without moral compunction in the present military establishments.

It is important to realize that the results of a single nuclear blast cannot be given with complete precision, as the variables of population density, building structure, topography and weather conditions all must be taken into account. This can be seen from the bombs dropped on Japan. The thirteen-kiloton atomic bomb dropped on Hiroshima killed approximately 130,000 people, while the twenty-one kiloton plutonium bomb dropped on Nagasaki killed less than half this amount, about 55,000. (These estimates are based upon the median between the official US figures and the official Japanese figures.)

The one megaton bombs that we are considering in the limited war scenario for Western Europe, while considered small by comparison with the strategic weapons in stock, are in fact about fifty times bigger than the Hiroshima bomb and will cause roughly the equivalent damage over an area about thirteen times as large. This is a difficult estimation to make because European housing and buildings are of much sturdier quality than those of World War II Japan; but on the other hand, modern European cities are much more densely populated and much more dependent upon complex centralized services. Bear-

Typical Measurements of Killing and Destroying Effects of a 1-Megaton Nuclear Bomb*

Effect	Range	Area
Overpressure, or blast:		
a. Crater (everything is turned to vapor within about the same range)	200 meters	12-50 ha
b. Destruction of apartment houses, with death of most inhabitants	6.0 km	100 km²
c. Destruction of wooden homes, with death of most inhabitants	9.0 km	250 km²
Nuclear radiation:		
d. Death due to prompt radiation	2.5 km	20 km²
e. Nuclear fallout sufficient to kill most persons in the open	elongated area reaching out to 100 km	± 150,000 ha
f. Nuclear fallout sufficient to kill persons sheltered in basements of houses or apartments	elongated area reaching out to 25-50 km	10,000-40,000 ha
g. Numbers of persons who die prematurely from leukemia and other forms of cancer in other parts of the world	worldwide	very roughly 1,000 persons per megaton
Other:		
h. Forest blown over by blast	7-10 km	15,000 ha
i. Dry forest ignited	10-20 km	30,000-120,000 ha

* 1 meter = 3.3 feet
1 kilometer (km) = 0.62 miles
1 square kilometer (km²) = 0.39 square miles
1 hectare (ha) = 2.5 acres

ing these factors in mind, it can be estimated that a one-megaton blast would kill about 1,000,000 people in a city of several million; in cities of smaller populations, the death-rate would simply be a major fraction of the total number of human lives present. The table on page 116 indicates typical measurements of the effects of a one megaton nuclear blast.[308]

If one were to take the populations of all the major Western European cities and then compute the number of one-megaton bombs needed to kill every single exposed person in all of Western Europe, the result is to be seen in the table following:[309]

Western Europe's Vulnerability to Annihilation by Fallout

Number of 1-megaton nuclear bombs needed to kill by delayed radiation (fallout) all persons in the open in Western Europe. *

Country	Population (million)	No. of Cities with Population over 200,000	Number of 1-megaton bombs necessary
Austria	7.5	3	56
Belgium	10	5	21
Denmark	5.1	2	29
Finland	4.9	3	225
France	51	20	365
FR Germany	67	30	166
Greece	8.9	2	88
Ireland	2.9	1	47
Italy	59	18	201
Luxembourg	0.35	–	2
Netherlands	15	11	23
Norway	4	2	216
Portugal	9.6	2	61
Spain	35	13	337
Sweden	8	3	300
Switzerland	6.2	5	27
United Kingdom	57	24	163
Total		144	2,327

*This hypothetical example for the number of persons that would be killed by *delayed* nuclear radiation assumes ground bursts and a distribution of bombs that would maximize fallout.

The number of persons that would be killed as a result of the energy released in the forms of *blast, thermal radiation* and *prompt* radiation as a result of one, 1-megaton explosion is estimated to be at least one million.

As can be seen, 2,327 one-megaton bombs are needed virtually to extinguish Western European life. There are in the Soviet arsenals alone, however, more than *3,500* tactical nuclear weapons deployed and programmed specifically for use in and against Western Europe. The Soviet strength can be seen from the next table:[310]

Soviet Tactical and Strategic Nuclear Systems

Type	Number	Range (miles)	Yield*
Tactical Systems			
1. Howitzer	100	25 km	few kt
2. Short-range, land-based missiles	1,000	6-800 km	10 kt-500 kt
3. Medium-range, land-based missiles	600	1,500-3,000 km	0.5 Mt-3.0 Mt
4. Short range, sea-based missiles	500	100-500 km	20 kt-500 kt
5. Nuclear-capable-tactical aircraft	2,400	600-3,000 km**	kilotons to megatons
Strategic Systems			
6. Strategic missiles, land- and sea-based	2,400	to 12,000 km	Mt (to 25 Mt)
7. Strategic aircraft	150	10,000 km (?)	Mt (to 25 Mt)

* kt = 1 kiloton or 1,000 tons
1 Mt = 1 megaton or 1,000 tons

** Strike radius

Countering these 3,500 Soviet weapons are *10,000* NATO tactical weapons.

I indicated at the outset that the scenario was that of a 'limited nuclear war', a favourite phrase in current nuclear war-games talk. It is clear, however, that only a 'limited' number of weapons on a 'limited' number of cities would be enough to kill virtually every exposed person from Scotland to Sicily and from Finland to Spain. After this destruction, both the US and the Soviet Union would still have the major portion of their nuclear arsenals left, presumably to be used against each other.

A recent study by the Congressional Office of Technology Assessment has in fact examined the effects of both limited and all-out nuclear wars between the Soviet Union and the United States. Released in June 1979, the study examines the possible scenario just considered for Western Europe: namely, that of one-megaton bombs being used against population centres. If Detroit was hit by a one-megaton bomb, for example, besides the million or so immediate deaths, there would be over 500,000 people injured. More than half of the 18,000 hospital beds would have been destroyed in the blast, however, and another 4,000 damaged, thus leaving only 5,000 beds to handle 'only 1 percent of the number injured'.[311] The study also points out that

Summary of Effects of Nuclear Attack

Case	Description	Main causes of civilian damage	Immediate deaths	Middle-term effects	Long-term effects
1	Attack on single city: Detroit and Leningrad: 1 weapon or 10 small weapons.	Blast, fire, & loss of infrastructure; fallout is elsewhere.	200,000-2,000,000	Many deaths from injuries; center of city difficult to rebuild.	Relatively minor.
2	Attack on oil refineries. limited to 10 missiles.	Blast, fire, secondary fires, fallout. Extensive economic problems from loss of refined petroleum.	1,000,000-5,000,000	Many deaths from injuries; great economic hardship for some years, particular problems for Soviet agriculture and for U.S. socio-economic organization.	Cancer deaths in millions only if attack involves surface bursts.
3	Counterforce attack; includes attack only on ICBM ships as a variant.	Some blast damage if bomber and missile submarine bases attacked.	1,000,000-20,000,000	Economic impact of deaths; possible large psychological impact.	Cancer deaths and genetic effects in millions; further millions of effects outside attacked countries.
4	Attack on range of military and economic targets using large fraction of existing arsenal.	Blast and fallout, subsequent economic disruption; possible lack of resources to support surviving population or economic recovery. Possible breakdown of social order. Possible incapacitating psychological trauma.	20,000,000-160,000,000	Enormous economic destruction and disruption. If immediate deaths are in low range, more tens of millions may die subsequently because economy is unable to support them. Major question about whether economic viability can be restored – key variables may be those of political and economic organization. Unpredictable psychological effects.	Cancer deaths and genetic damage in the millions; relatively insignificant in attacked areas, but quite significant elsewhere in the world. Possibility of ecological damage.

burn victims would number in the tens of thousands while in the entire country there are no more than 2,000 beds in specialized burn centres.

Even in a limited nuclear war aimed at each other's nuclear bases and missile silos, generally kept well away from population centres, the study finds that 'while the consequences might be endurable, the number of deaths might be as high as twenty million'. In the ensuing months after this 'limited' war, another twenty million could be expected to die of radiation sickness, starvation, lack of shelter and limited medical supplies. It was also pointed out that 'The long-term ecological damage cannot be excluded . . . some regions might be almost uninhabitable.'

Should this 'limited' nuclear war escalate into a full-scale nuclear exchange, 165 million of 220 million people would be immediately killed in America, the study estimates, and in the Soviet Union about 100 million of 261 million people would die. Survivors in both countries would live in conditions that would be 'the economic equivalent of the Middle Ages'. The chart on page 119 details the damage that can be expected given different nuclear exchanges.

N. THE AGE OF OVERKILL: AN ILLNESS OF POWER

I pointed out that the psychological impact of the atomic bomb upon both the Hiroshima and Nagasaki survivors as well as upon all those involved in dropping the bomb was that of a sudden and traumatic shift from normal existence to death-dominated life. On 6 August 1945, the world was transfigured, and victims and victimizers alike underwent the experience of death immersion, an experience that has remained engraved on the collective consciousness of humankind ever since. In reacting to this death immersion, most survivors psychically closed themselves off, thus beginning a psychological dynamic in their minds that was ultimately to culminate in paranoia and an identification with atomic weapons and the death they symbolize. In discussing paranoia, I pointed out what Elias Canetti observes, that it 'is an illness of power in the most literal sense of the words'. It is an illness that seeks power not only over life but over death, but because it is involved with an identification with the very death it is attempting to conquer, it can use no other weapon to fight death than the threat of death. Therefore paranoia ultimately results in destructive behaviour to others as well as to the self. While paranoia ordinarily seeks to divert death to others in order to spare the one who is paranoid, Canetti makes the point that 'once (a person) feels himself threatened, his passionate desire to see *everyone* lying dead before him can scarcely be mastered by his reason'.[312]

This is a most profound insight, for it goes a long way towards explaining why it is that out of mutual mistrust and paranoia for each other, the Russians and the Americans have built up such huge nuclear arsenals that they now threaten every single person on the planet with death immersion. The current world situation, therefore, with each human being not threatened merely with one death but with arsenals that can destroy each man, woman and child *scores* of times over – this situation should be understood as acute paranoia, something that should be dealt with before it culminates in the double suicide that acute paranoia makes inevitable if left untreated.

This contemporary paranoia, however, should not be seen as something that has arisen since Hiroshima; rather, Hiroshima and the Age of Overkill should be seen as a culmination of a classical system of world politics, the core of which has been and still is the exercise of power.

Max Lerner is very helpful in understanding this point. He lists ten premises and principles of classical world politics which reveal that inherent in the *structural* way in which the world has ordered itself in recent centuries is a *psychological* perspective that has produced the Age of Overkill.[313]

1. We are within a system, he says, that is based on power and the uses of power – for supremacy of one nation over the other, for war and the maintenance of peace, for equilibrium within nations and for the changes of rulership both in and between nations. Power 'is the prime mover, and it is also the goal toward which everything in the system gravitates'.

2. Since the ruling passion is for power, the underlying assumption of classical world politics is that of the *scarcity of power*. In this sense, classical politics is similar to the classical *laissez-faire* economics of capitalism, whose underlying premise was that there was a scarcity of wealth and resources so each person should strive for the largest possible share of what is available.

3. Because of the premise of the scarcity of power, classical politics, like classical capitalism, is competitive and aggressively

hostile. The aim of each nation is not only to acquire more power but actively to prevent its enemies from acquiring any. Thus the premise of scarcity yields the corollary premise of the *enemy*. The world becomes an arena of enemies and potential enemies and allies and potential allies: hence, the concept of alliances between nations, the art of diplomacy, and the utilization of war as an instrument of foreign policy.

4. Given the constant awareness of an enemy competing for power, classical politics can be characterized psychologically as a system predicated upon *suspicion and fear*, along with the secrecy and mistrust such attitudes imply. The characteristic ethos is therefore that of the amoral reliance upon the 'reason of state' or 'national security' to legitimate whatever actions a particular nation feels it must take to protect itself.

5. In the competition for power, classical politics puts value upon what Lerner calls 'sinews of strength': territory, natural resources, a population sufficient for an army and labour, a favourable balance of trade and payments, and the acquisition of wealth and strength. The emphasis is upon the tangible and the observable, not upon the intangibles of ideals and morality. The major exception to this is the cultivation among the young in particular of patriotism and national honour as ideals that one should be prepared to sacrifice life for.

6. Fundamental to classical world politics is the notion of *nation-states*. Anything less has no substantive value because it has little to show for itself in terms of the accepted criteria of what constitutes sinews of power. Hence the constant pressure in recent times towards nationalistic movements and the carving out of new nations. Once this has been achieved, the drive of the new nation is not towards either internal freedom or external co-operation but towards self-sufficiency economically and militarily. All else is generally subordinated to this task, for within self-sufficiency is independence and national security.

7. Fundamental to the notion of nation-states is the notion of *national sovereignty*, meaning that each nation refuses to acknowledge a source of authority outside itself. Any participation in treaty or multi-national organization is only done if it accords with perceived self-interest.

8. Closely connected with national sovereignty is *national interest*, an internal dynamic within each nation that seeks out of the competitive and suspicion-filled chaos in which it finds itself to discern what it needs and how best to obtain it. Lord

Palmerston made the remark once during a Parliamentary debate over whether England and Turkey should form an alliance that 'England has no permanent friends, England has no permanent enemies, England has only permanent interests.'[314] This statement forms the essence of classical world politics.

9. As each nation pursues its own self-interest it must somehow attempt to ensure that it will not be destroyed by its competitors. This Lerner describes as the 'balance-of-power principle'. This means that when any nation gets too powerful so that it threatens its enemies with possible extinction or subjugation, these enemies must attempt in some way to restore the balance. Thus classical politics is a system based on a notion of *equilibrium*. To ensure this equilibrium the ability to make war is essential, for only by being able to inflict unacceptable damage to anyone perceived as a threat does any nation feel secure enough to continue its competition for the limited amounts of resources and power the other nation is upsetting the equilibrium in order to get.

10. *War* is thus the ultimate referent of classical politics. It is the grounding principle that ensures that the enemy in the pursuit of power can be either neutralized or defeated; it is the basis upon which the suspicion and fear rest; it is the defence behind which nation states continue existing and competing.

Power politics – *Machtpolitik* – is therefore synonymous with classical politics as there is no way of restoring the balance of power except through the use of war; this because each nation refuses to recognize any referent outside itself.

The Age of Overkill has brought the classical system of world politics to an end, however, because while the first principle of the classical world was that of the *scarcity* of power, the reality of nuclear arsenals is that of a *surplus* of power. The tragic paradox is that while using weapons no longer suitable to the classical political system, the nation-states with nuclear arsenals are using these weapons to defend interests and to maintain the paranoia *produced* by the classical system. And yet the classical political system is so deeply engrained that even while knowing that the entire planet is being threatened, the nations cannot stop their escalation to develop more weapons, more sophisticated guidance systems, more accurate missiles.

If the cause of this current self-destructiveness was merely the paranoia produced by a particular political methodology, there would perhaps be hope of changing it, of replacing the classical system with one more modern, more condusive to inter-group and inter-nation co-operation. But while the present nuclear arsenals are manifestations of the classical system of world politics, it must be noted that classical world politics is merely the manifestation of something deeply engrained in the human being itself; namely, the aggressive and predatory nature of human instinct.

This point needs to be explained because it is fundamental. It is well known that human ancestry is firmly rooted in animal origins – this is the conclusion of several centuries of anthropological research into the evolutionary history of our species. What has been emerging with an almost equal clarity has been the *predatory* nature of our origin. This is to say, that since our emergence from the apes we have been predators whose natural instinct has been to kill with a weapon.

Remembering the different characteristics of the classical political system, it is interesting to note the different instincts we have inherited from our primate ancestors. Robert Ardrey, in his book, *African Genesis*, notes that these include the establishment and defence of territories; an attitude of perpetual hostility for the neighbouring territorial holder; the formation of social bands as the principle means of survival; an attitude of amity and loyalty for those within one's own group and mistrust and suspicion for those in other groups; and a varying but universal hierarchical system to establish dominance and submission within the group to ensure both the efficiency of the group and the promotion of natural selection.[315] Upon this deeply buried multifarious primate instinctual complex was added the necessities and opportunities of the hunt.

Ardrey exemplifies this through illustration of the lion. The non-aggressive primate, he says, rarely dies in defence of its territory; however, among the causes of lion mortality tabulated in the Kruger reserve of South Africa, death from territorial conflict is the second largest contributing factor. The non-aggressive primate seldom suffers much beyond humiliation in its quarrels for dominance; the lion, on the other hand, dies of such conflicts more than from any other cause. The forest primate suppresses many individual demands in the interest of the collective but still retains a measure of individuality; no collective in the animal world, however, can compare with either the organization or the discipline of the lion pride or the hunting pack of the wolf.

We can only presume, observes Ardrey,

> that when the necessities of the hunting life encountered the basic primate instincts, then all were intensified. Conflicts became lethal, territorial arguments minor wars. The social band as a hunting and defensive unit became harsher in its code whether of amity or enmity. The dominant became more dominant, the subordinate more disciplined. Overshadowing all other qualitative changes, however, was the coming of the aggressive imperative. The creature who had once killed only through circumstance killed now for a living.[316]

In short, building upon anthropological evidence generally, particularly upon the evidence uncovered by Leakey in the Olduvai Gorge and Dart in southern Africa,[317] Ardrey asserts that the human species 'originated as the most sophisticated predator the world has ever known'. War, he says, 'has been the most natural mode of human expression since the beginning of recorded history, and the improvement of the weapon has been man's principle preoccupation since Bed Two in the Olduvai Gorge'.[318]

Robert Bigelow, in his book, *The Dawn warriors: Man's Evolution Towards Peace*, amplifies this last point in a way that elucidates both the deep-seatedness of our aggressive drives but also the close connection as well between our feelings of co-operation and group support and our feelings of hostility and inter-group rivalry so characteristic of the classical political system. He begins his book, appropriately enough, with an image of the very phenomenon we are considering:

> A hydrogen bomb is an example of mankind's enormous capacity for friendly co-operation. Its construction requires an intricate network of human teams, all working with single-minded devotion toward a common goal. Let us pause and savour the glow of self-congratulation we deserve for belonging to such an intelligent and sociable species. But without an equally high potential for ferocity, no hydrogen bomb ever *would* be built. Perhaps our co-operation has something to do with our ferocity.[319]

Bigelow offers the thesis that these two apparently contradictory aspects of human nature are in fact complementary rather than contradictory. He points out that early human societies were organized in socially co-operative groups, much as the upper primates such as the baboon are today, but on a much higher level of complexity. The co-operation was confined, however, almost exclusively to individuals who belonged to the same group. Co-operation between groups was essentially non-existent, and the individuals of each group regarded all other groups as foreign and therefore as potential threats to their survival.

Effective group action necessitated effective communication to facilitate co-operation. This, in turn, required developed brains. Those groups with the most highly evolved brains and consequently the greatest capacity for effective communication and co-operation in attack and/or defence, maintained themselves longest in the areas most condusive to their growth. They were less susceptible to defeat and therefore less often driven into sparse areas to starve or be preyed upon. On the average, therefore, the groups with the most effective brains would prevail. This understanding, says Bigelow, offers 'an explanation of human evolution that considers warfare as the force that demanded a threefold increase in the size of the brain in a mere two or three million years'.[320]

The ingredients of evolution, genetic variability and natural selection determined that survival went to that new combination most able to learn social co-operation for defence and offence against threats of would-be predators – both human and non-human. Those groups failing to co-operate effectively were slaughtered or driven into less hospitable lands.

Social co-operation, then, was demanded by the ferocity of human competition for scarce resources: 'Co-operation was not substituted for conflict. Co-operation-for-conflict, considered as a single hyphenated word, was demanded – for sheer survival.'[321]

As the human brain developed further, the advantage of *inter-group* co-operation was seen, a difficult step to take in that it required co-operation with previous enemies and foreigners. As soon as the first two human groups formed a military alliance against a third party, however, the single group was faced with grave peril. Bigelow asserts this inter-group co-operation-for-conflict was a central factor in quickening the pace of human evolution.

Co-operation-for-conflict at the inter-group level had the following results: the contending armies became larger, the weapons more devastating the casualties higher. But, and this is of critical importance, the areas within which relative peace prevailed also increased. *The larger the group, therefore, the larger the area of internal peace*, a point that will be further explored later in discussing a possible solution to the present crisis. The point being stressed now, is that as history has progressed the restricted areas of inter-group co-operation-for-conflict have gradually expanded. Humans have gone from hunter-gatherer groups, to tribes, to city states, to kingdoms, empires, until today whole blocs of nations exist in alliances such as the Common Market or NATO or the Organization of African Unity. War has simultaneously become bloodier and more destructive. We are thus today co-operating as never before but we are also divided and competing as never before, until all sides have armed themselves to the point of being able to blow the entire planet into oblivion.

Indeed, this admixture of co-operation-for-conflict has produced such a permanent psychological state of paranoia in our day, such a global competition for more armaments to add that little bit more security we deem essential to our national defence, that we have entered what Lerner calls a 'strange new world' of needs and fears arising from primal instincts and classical political conditioning and yet with nuclear weapons that have made the *satisfaction* of those needs and fears in the manner we have become accustomed over the millennia now obsolete. Consider, he asks, the world we have created for ourselves:

> a world of buttons existing to be pressed at need, with the underlying premise that they will not have to be pressed if one threatens clearly enough to press them; a world of peace to be achieved through the promise of nuclear terror; a world bedecked with the new vocabulary of 'credible first-strike capability', 'second-strike (retaliatory) capability', 'escalation', 'reciprocal fear of surprise attack', 'calculated preemption', and even 'striking second-first'; a world in which chance has all but been squeezed out, through 'fail-safe' built-in controls of the weapons systems, with crucial decisions reserved for a few men at the top; a world of constant wariness and suspicion, of a death's-head game in which each side is bent on convincing the other that it has the means and the will to destroy him utterly before it is itself utterly destroyed, that it is powerful enough to instil terror, yet pre-

> pared enough to feel none. Only thus, it is argued, will the weapons so arduously conceived, placed on the drawing board and brought into the stockpiles remain unused. Out of this nettle, deterrence, we presumably pluck this flower, safety.[322]

It should of course be pointed out that none of the above suggestions should be considered as being given as a complete explanation of the Age of Overkill. The problem is as multifarious as the different dimensions of our lives that it impacts. There are economic reasons, political causes, deep-seated psychological attitudes and primal instinctual forces at play. All cohere relationally to produce the imbalance we are currently being forced to experience, an imbalance that could mean the extinction of our species or at least the termination of civilization as we know it.

The point that I wish to stress here is that this imbalance is indeed real and that it is a dynamic that includes *all of us*. No one can remain apart, no one can think that they are somehow exempt, for the threat is a global one, drawing all of humanity into its field of influence. As I pointed out, the paranoia felt by the super-powers is such that they insist that every person's life be placed in equal jeopardy to their own.

The tragedy of this imbalance in which we all find ourselves is that it is not limited to weapons spirals, whether nuclear or conventional, as the different blocs of power co-operate in order to compete better. The Atomic Age has permeated other sectors of human existence, primarily that of energy resources and consumption. Indeed, the further proliferation of nuclear weapons is inseparably connected with the growing number of nuclear reactors around the world, now in over forty-three different countries. In order to understand more fully the dimensions of the current problem, therefore, it is necessary to examine the function and implications of nuclear *reactors*, even as we have examined that of nuclear *weapons*. It is only once the enormity of the problem that confronts us is understood that we will be in a position to begin an articulation of any possible solution. From Hiroshima, therefore, we must go on to Harrisburg.

II

The Road to Harrisburg

The connection between atoms for peace and atoms for war has already been discussed, as the plutonium waste for nuclear reactors can be used in the construction of nuclear weapons. What needs to be further explored, however, are the nuclear reactors themselves, for they are proving to be just as dangerous in their own right as nuclear weapons.

While the impact of nuclear weapons has been largely that of producing extreme trauma in the *psychological* realms of humanity, even though their purpose is to inflict bodily and property damage – this because they have only been used on two occasions – the impact of nuclear reactors, while not yielding psychological impact or even meeting much public opposition when they began to be installed, have had a far more immediate impact, particularly on peoples' *bodies*. The psychological impact of nuclear weapons, therefore, and all that entails in terms of death immersion, guilt, psychic numbing, paranoia and death identification is balanced in almost a complementary way by the impact nuclear reactors have had on the human body, in terms of producing cancers, leukemia, stillbirths, genetic mutations, and environmental pollution.

In order to understand this more fully, it is necessary to discuss the common denominator between nuclear weapons and nuclear reactors: *radiation*.

A. RADIATION

What caused the atomic explosion at Hiroshima was a chain reaction of neutrons released from the uranium-235 inside the bomb. Called *fission*, this involved splitting a uranium atom to release its neutrons, which in turn split other uranium atoms, which in turn split others – thus causing the desired explosion which in the first one millionth of a second caused the half-mile wide fireball. The chain reaction was virtually instantaneous. Such fissioning of uranium-235 atoms is also the basic principle in the operating of nuclear reactors, although instead of inducing the atoms to fission in an uncontrolled instantaneous way, the chain reaction is *controlled and sustained* over a period of time. A sustained chain reaction means that each neutron lost by causing the fissioning of the next uranium atom is replaced by exactly one more neutron. The system therefore, has a very small fissioning process going on, but one that can be kept up indefinitely. When a reactor has been started up and the fissioning process has become a sustained chain reaction, it is said to have attained *criticality*.

What is important to know is that for both atomic weapons and nuclear reactors, whether operating by fission or the reverse process of fusion – collapsing atoms, fusing them into one another rather than splitting them – the effect is the same: *the massive release of radioactivity*.

This last point can be better understood through illustration. When a stray neutron from one uranium-235 atom hits the nucleus of another uranium-235 atom the result is the rupture, the fissioning of the second atom. What happens when this second uranium-235 atom splits is that about 40% of its nucleus flies off in one direction and 60% in the other. This bursting of the nucleus happens with so much energy that when the masses of the atom are later tallied, there is a shortage. This is because some of the *mass* of the original nucleus has been converted into *energy*. This transformation is the source of all the energy released in nuclear events. When this happens suddenly to many uranium atoms at the same time the result is a bomb; when this process is sustained and controlled, the result is what happens inside a nuclear reactor.

The fissioning of a nucleus is the most violent type of breakdown a nucleus can experience. There are other ways it can be broken down, however. Under stress, uranium-238 may release two protons and two neutrons, identical in all respects to a helium atom. These protons and neutrons are known as an *alpha particle*, which, when released, comes flying out of the uranium atom at such a high speed that it ploughs a furrow through whatever it hits. A more delicate form of breakdown involves the emission of an electron which, because it is emitted at such an extremely high speed, is called a *beta particle*. One final form of emission can occur and involves the nucleus squirting out a burst of energy closely akin to light, although much more energetic and invisible. This is known as a *gamma ray*. A gamma ray is identical to an X-ray, except that it comes from the nucleus of the atom while the X-ray comes from the electron layers around the nucleus.

An atomic nucleus can thus alter itself in four ways: by fissioning, by emitting alpha or beta particles, or by emitting gamma rays. Any atom that contains an unstable nucleus from which these particles and rays can shoot out is known to be *radioactive*. The emissions – the neutrons, alpha and beta particles, the gamma rays – are known as *radiation*. If a nucleus emits 37,000,000,000 such emissions per second, it is said to have emitted one *curie* of radioactivity.

The radiation that is emitted travels until it hits something that will stop it. The consequences of this interchange will depend upon the substance that is hit and on the type of radiation involved. For example, an alpha particle, the largest of the four types of radioactive emissions, interacts dramatically with any atomic structure it hits, dislocating electrons and knocking the nucleus out of place, but because it has such an enormous impact it quickly uses up its energy. Once emitted, therefore, an alpha particle can be stopped by an ordinary sheet of paper. A beta particle is much less destructive, dislodging only other electrons, and can as a result travel further, requiring the thickness of something like a thin plate of metal to use up all its energy. A gamma ray can travel even further, causing a relatively small amount of damage at any given point along its path. A neutron likewise travels great distances and is generally only slowed down by direct impact with the nucleus of another atom. Both gamma rays and neutrons can penetrate concrete more than a metre thick.

If any of these four radioactive emissions dislodges an electron from an atom, the atom becomes an *ion*. Radiation from such atomic nucleii is called *ionizing radiation*. When ionizing radiation passes through a substance it causes *structural* changes – sometimes permanent, sometimes temporary, most always damaging. The more ionizing radiation impacts a substance, the greater the disruption and possible damage. The basic unit of radiation exposure is called a *roentgen*, after the discoverer of X-rays, Wilhelm Roentgen.

The effects of ionizing radiation are particularly damaging if the substance it hits is that of *living matter*. This is because the delicate molecular arrangements that make up the human body, for example, are easily unbalanced by radiation. The units used to measure the radiation impacting the human body and other living things are the *rad* (radiation absorbed dose) and the *rem* (radiation equivalent man). For beta and gamma radiation one rad equals one rem. For neutron and alpha particles, however, one rad may equal up to twenty rem, depending on the energy of the particles. Rem is the most important term here, for it measures the actual amount of damage done to the body.

One further term that is important to know is *half-life*. This refers to how long a radioactive particle or ray can be expected to remain active. If there are 100 strontium-90 atoms, for example, it will take twenty-eight years for half of their nucleii to have decayed to the point of there being 50 nucleii left. After another twenty-eight years there will be a further decay of 25. Every twenty-eight years half of what was originally there will have decayed away. Thus the half-life of strontium is twenty-eight years. Each radioactive particle or ray has a half life. Some last for only fractions of a second; others last for millions of years.

The important thing to understand from the above discussion is that when something like a uranium-235 atom is fissioned, the resulting radia-release causes *structural damage* in whatever it hits, particularly if what it hits is *living tissue*. Further, it is important to bear in mind that the radioactivity emitted is not necessarily used up quickly and forgotten. Many particles have extremely long half-lives, which means that they remain radioactive, damaging whatever molecular or atomic structure they impact, for a long time.

This fact can be illustrated by examining four of the many radioactive agents that are emitted each day in the routine operations of a nuclear reactor: strontium-90, cesium-137, iodine-131 and plutonium-239.

Strontium-90 and cesium-137 are both found within the reactor fuel rods under normal operating conditions. Both emit penetrating radiation and large amounts of heat and are so toxic that they must be isolated from the environment for more than a century. Strontium-90 is a bone seeking isotope, meaning that if taken into the human body it will accumulate in bone marrow. This is because its atomic structure is much like that of calcium, so the body does not attempt to excrete or combat it. Yet it is so lethal that to dilute *one gallon* of liquid containing 50 to 100 curies of strontium to legal levels would require from *500,000,000,000 to 1,000,000,000,000 gallons* of water. The effect of strontium-90 on the body is to produce leukemia.

Cesium-137, like strontium-90, remains radioactive for hundreds of years and is equally lethal. It emits gamma rays and concentrates in the muscle tissue and in the ova of females, producing cancer.

Iodine-131, while as toxic as both cesium and strontium, only has a half-life of eight days. If ingested, however, it locates itself generally in the thyroid gland, causing cancer.

Plutonium-239, of all the twenty-seven radioactive particles and rays emitted during the normal fissioning process of nuclear reactors, is the most lethal. We have already discussed how it can be utilized in the construction of nuclear bombs; indeed, it is the essential ingredient in their construction. What needs to be pointed out here is its lethal effects on the body. In terms of the human body, plutonium is quite simply the most lethal carcinogen known.[323]

Single particles of plutonium weighing *one millionth of a gram*, so small they can only be seen under microscopes, can cause cancer of the lung if ingested. The toxicity of plutonium can only be understood by comparison. Potassium cyanide, pellets of which are used in gas chambers to kill convicted criminals within minutes is generally thought of as being one of the most toxic substances available for use in capital punishment conditions. The single particle of plutonium just described, however is *20,000 times* as deadly as a pellet of potassium cyanide. According to Dr John Gofman, co-discoverer of uranium-233 and three other radionuclides and Assistant Director of the Lawrence Radiation Laboratory in Berkeley, California, from 1963 to 1969, a mere .011 micrograms of reactor plutonium is enough to cause cancer in a cigarette-smoking male. This .011 microgram equals 385 *trillionths* of an ounce. For non-smokers, the amount needed for cancer is 1.4 micrograms. One can understand this amount by being aware that in each pound of reactor plutonium there are *338,000,000* cancer doses for non-smokers. If one took just ten pounds of plutonium, therefore, and distributed it equally to each person around the world, all would have a high probability of dying of cancer.[324]

Nor is this all. Besides causing cancer, plutonium is also attracted to the human reproductive system where it lodges in the genetic coding, causing serious damage in the reproductive chromosomes. This causes genetic mutations in the next generation. Anyone ingesting plutonium, therefore, may not only come down with cancer of the lungs but will have a higher-than-normal probability of having genetically mutated off-spring.

Much more toxic than other radioactive agents, plutonium also lasts much longer than most – its half life is 28,000 years, meaning that it is potentially dangerous to human life for nearly 500,000 years. It is so volatile that small fragments of plutonium burn spontaneously on exposure to air, producing an almost invisible smoke. With its long life, it can be scattered into the air or soil and be ingested into one living being after another for thousands of years, mutating each one as it impacts them and causing cancer and genetic defects in succeeding generations.

Plutonium-239 is perhaps so lethal and long-lasting because it is not found in nature. Plutonium-239 is *human-made* and was first produced in the American attempt to construct an atomic bomb. When the scientists of the Manhattan Project saw its qualities they named it after the Greek god Pluto. Pluto is the god of hell.

One would think that because only one-millionth of a gram of plutonium is enough to cause cancer and/or genetic mutations; that because only ten pounds of it are enough to construct a nuclear bomb that could pulverize a city; and that because plutonium must be somehow contained so that it does not get out into the human or natural environment for nearly a half-million years, an attempt would be made by the nuclear industry and the government to restrict its production. This is not the case, however. Each operating nuclear reactor produces between *400 and 600 pounds* of plutonium each year in its normal operations. So much is being produced around the world that in 1972, the AEC projected that by the year 2000 the American reactors alone will have produced *50,000 tons* of it.

This situation becomes tragic when one realizes that plutonium, like all radio-activity with the exception of krypton, does not disperse itself evenly throughout the environment. Rather, it tends to concentrate itself in the *organic life* in the environment and in the *food chain*. If cesium-137 was released into a river, for example, the fish in that river would absorb 1,000 times as much cesium as would be found in the river as a whole. If strontium-90 was released into a pasture, the cows and their milk would have a much higher concentration of the strontium than the grass or the ground would. This concentrated radioactivity is then passed on to anyone eating the fish or drinking the milk. The ways in which radiation finds its way into organic life and into human beings can be seen from the following diagrams:

RADIATION EXPOSURE PATHWAYS

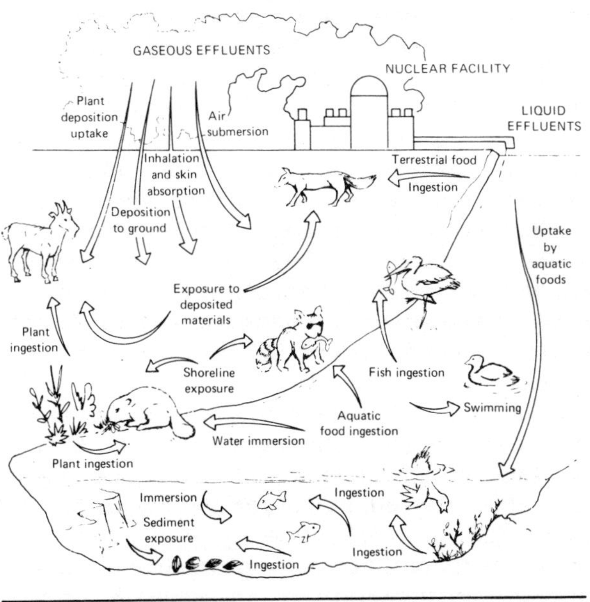

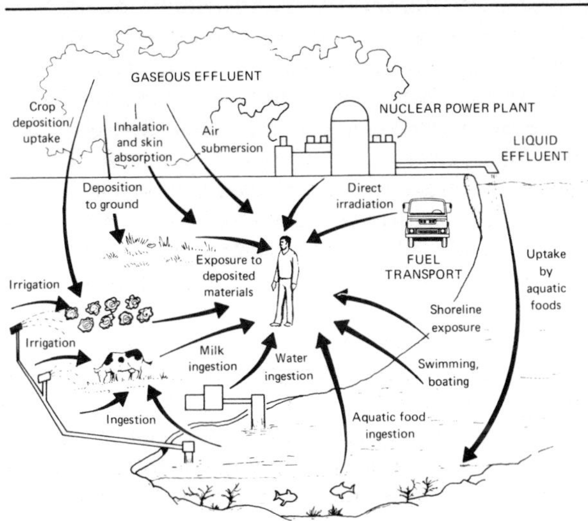

The corollary principle to the tendency of radiation to concentrate in organic life, is that the *younger* that life is the more vulnerable it is to the impact of the radiation. Children are much more susceptible than adults; infants and babies still in the womb are the most susceptible of all. This can be illustrated by recalling the experience of the mothers in Nagasaki after the bombing. Not one single baby *in utero* at the time of the blast whose mother was within 3,000 metres of Ground Zero survived. Not only does radioactivity concentrate itself in the food chain and in organic life, therefore, it attacks that part of life that is the most vulnerable: the young.

Quite apart from the effects of radiation upon the body in terms of increased cancer and leukemia rates, however, is a growing body of evidence indicating that radiation produces a generalized ageing effect on the body. This is the conclusion of the research of Dr Rosalie Bertell of the Roswell Park Memorial Institute after nine years of research examining a population base of 13 million people in the states of New York, Maryland, and Minnesota. Called the Tri-state study, it is perhaps the largest leukemia survey ever done. The conclusion Dr Bertell has drawn is that 'The health effects connected with radiation seem to be a *secondary* phenomenon, the primary one being the acceleration of the breakdown of the body, which is the ageing process.'[325]

Dr Bertell's data confirm that exposure to one rad of radiation (the amount of radiation one receives from a heavy abdominal or spinal X-ray), is the equivalent to one year of ageing. This is not so say that one will necessarily die one year sooner; rather, it means that one increases one's probability of coming down with old-age diseases such as cancer; leukemia, heart diseases, diabetes, even allergies at an earlier age than normal. The *quality* of one's life is therefore being adversely affected by radiation exposure, not necessarily the quantity of life.

What is important to know in terms of the radiation releases coming from nuclear facilities is that according to the existing standards, the general public can legally receive the equivalent of seventeen chest X-rays per year without the power plants exceeding 'acceptable' limits. Workers can legally receive several hundred.

The amount of radioactivity being produced in each large nuclear reactor has been calculated by John Holren, a physicist at the University of California. He estimates that in each reactor there are approximately 12,000,000,000 curies of fission products, 3,000,000,000 curies of actinides (those elements listed after Actinium on the Periodic Table of Elements) and 10,000,000 curies of activation products (elements made radioactive by neutrons).[326] What this all means in concrete terms is that *each large nuclear reactor of 1000 megawatt capacity contains as much radioactivity as would be released if one thousand Hiroshima bombs were simultaneously exploded*.

Because of the extreme *toxicity* of this radioactivity, therefore, as well as because of the *amount* of radioactivity involved, each reactor must be constructed so as to be able to contain it all within itself without leaking any out into the environment. This is done by keeping the reactor core, where the fissioning of the uranium takes place, constantly surrounded with either water or gas, depending on the type of reactor. Over both the reactor core and its surrounding coolant is a heavy metal covering called the *reactor vessel*. Around this system is another even stronger *containment structure*. How this system works can be seen in the diagrams on pages 138-39.[327]

The surrounding coolant not only cools the reactor core but transmits the heat produced by the fissioning process to turn the blades of a turbine located alongside the pressure vessel, thus creating electricity by rotating a generator. This can be seen from the following example of a pressurized water reactor (PWR), the most common reactor type:[328]

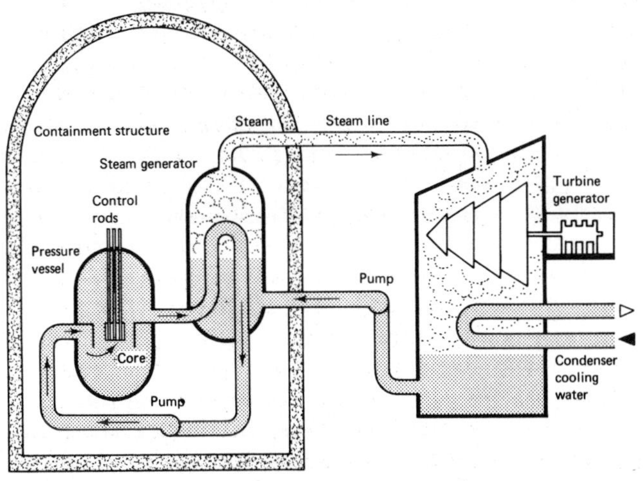

Pressurized Water Reactor (PWR)

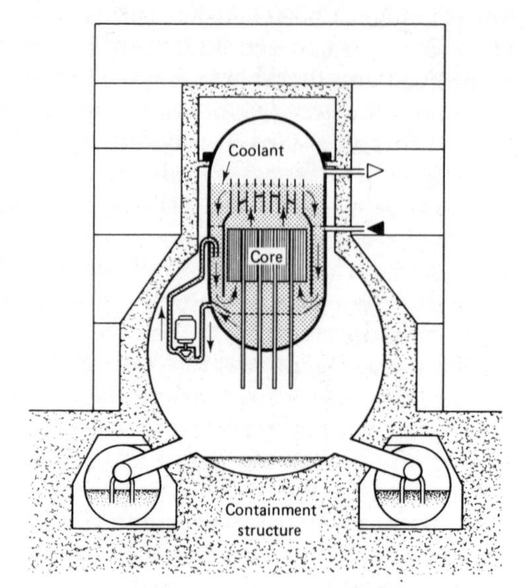

Boiling Water Reactor (BWR)

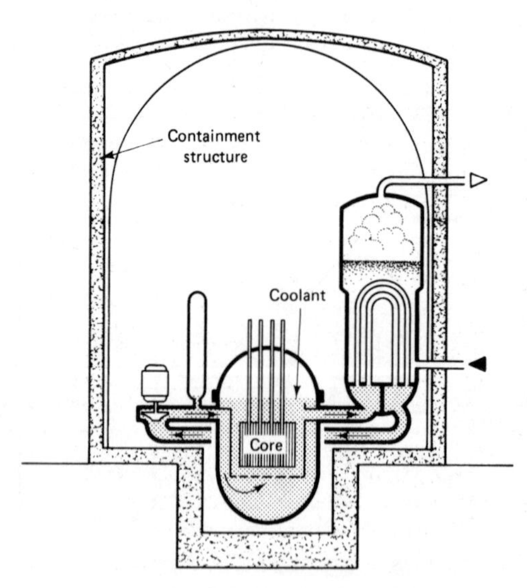

Pressurized Water Reactor (PWR)

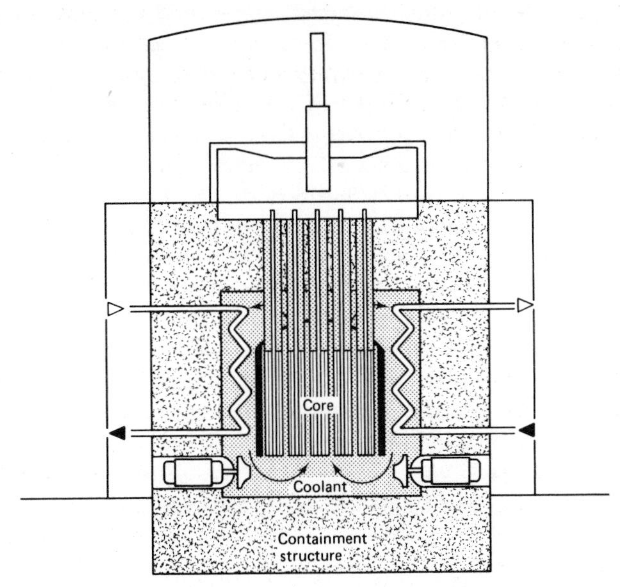

Advanced Gas-cooled Reactor (AGR)

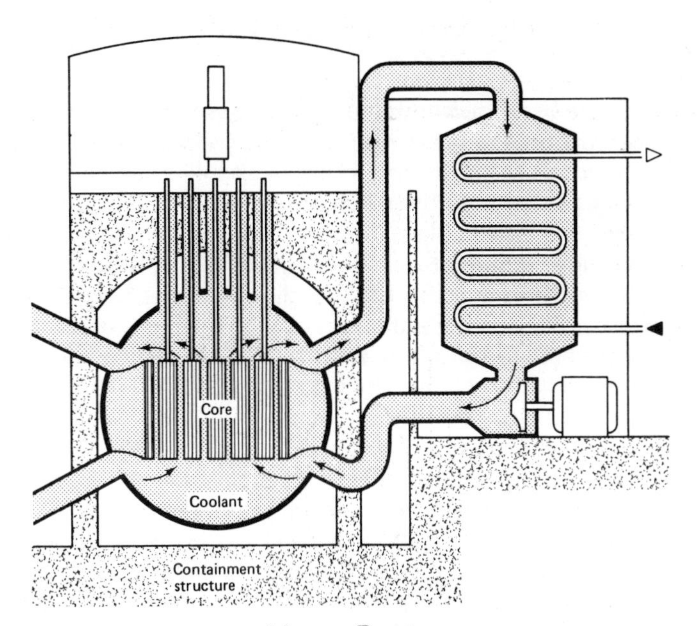

Magnox Reactor

As can be seen, once the heat is generated its transformation into electricity is essentially the same as what is involved in a conventional power plant using coal, oil or gas to produce the initial heat. The vast difference between a nuclear reactor and the conventional plants, however, is in the enormous amounts of radioactivity produced by the reactors. Should there by any rupture in the containment barriers due to a malfunction in the reactor system, radiation would be released. Remembering that each 1,000 megawatt nuclear reactor contains as much radioactivity in its reactor vessel as 1,000 Hiroshima bombs, a release of even a fraction of that amount is enough to do serious environmental and human damage. There is no explosion like the Hiroshima bomb, just the release of radioactive particles and rays which are invisible, tasteless and odourless. Their properties can only be felt after impact.

B. MELTDOWN EFFECTS AND PROBABILITIES

The most serious accident that can occur at a nuclear reactor involves the *loss of coolant* in the reactor vessel. If this occurs, and the back-up Emergency Core Cooling System fails to work as well, the result is a *meltdown*. What happens is basically simple: the coolant surrounds the reactor core, where the fissioning of the uranium is taking place, in order to keep it at a constant and relatively low temperature. If this coolant is released in some way, either through a rupture in a pipe or in the reactor vessel, the nuclear fuel in the core heats up to such a degree that it melts together, forming a mass of white-hot radioactivity. It is conceivable that it will melt its way through the floor and foundations and begin burrowing its way into the earth. This is known as the 'China Syndrome'. During the meltdown, however, there would be gas and steam explosions serious enough to rupture the outside containment vessel, thus allowing enormous amounts of radioactivity to be released directly into the atmosphere. Given what has already been discussed about certain of the radioactive particles and rays inside each reactor, their release in massive quantities would be catastrophic.

In 1957, the US Atomic Energy Commission (AEC) contracted the Brookhaven National Laboratory in New York to research what would occur if there was a meltdown at one of its reactors. The result was the Brookhaven Report, entitled 'Theoretical Possibilities and Consequences of Major Accidents in Large Reactors'.[329] The study assumed that the reactor was 200 megawatts in capacity, located thirty miles from a city of one million, and involved in an accident in which only 50% of its radioactivity escaped. Given this scenario, the report estimated that there would be 3,400 immediate deaths, 43,000 injuries and up to $7,000,000,000 in property damage. The report did not estimate long term deaths or the number of genetic defects.

Another study done for the AEC in the same year was carried out by the Engineering Research Institute of the University of Michigan at Ann Arbor. It was released in July 1957, and was entitled 'Possible Effects on the Surrounding Population of an Assumed Release of Fission Products into the Atmosphere from a 300 Megawatt Nuclear Reactor Located at Lagoona Beach, Michigan'.[330] The study indicated that 133,000 people would receive radiation dosages of 450 rads and that half of them would die. It estimated that as many as 181,000 would be injured, meaning that they would receive dosages of at least 150 rads of radiation. Of the injured, many were expected to die. Upon receiving this report, the AEC quietly buried it in its files, much preferring to release to the public the more conservative estimates of the Brookhaven Report.

When the insurance companies asked to insure the nuclear reactors being built realized how much damage they could do, they refused to insure the utility companies building the reactors. Instead of pausing to reflect on whether this meant that perhaps reactors were too dangerous to the public, the AEC and the Eisenhower administration persuaded Congress to pass the Price-Anderson Act as an amendment to the Atomic Energy Act of 1954. While the Brookhaven Study had indicated that property damage alone might exceed $7,000,000,000, to say nothing about damages to the families of persons dead or injured by the accident, the Price-Anderson Act arbitrarily stated that the utilities would have to pay no more than $560,000,000. But even this amount was considered too much by the insurance companies; instead, they formed an insurance pool of $110,000,000 to cover any accident the utility companies might have. The result was that Congress, *using tax payers' money*, authorized the

AEC to make up the $450,000,000 difference. Besides not being covered for even a fraction of the amount of damage that will ensue if there is a major accident, therefore, the American public is even being forced to pay for whatever insurance it will receive.

In other countries the response of the government to a possible worst accident has been similar. In Britain, for example, the 1959 Nuclear Installations Act, which created the Inspectorate of Nuclear Installations responsible for the safety of nuclear power stations and research centres, also severely limited third-party liability in the case of a nuclear reactor accident. While continuing to assure the public that nuclear power is safe, therefore, neither governments nor nuclear industry nor insurance companies have been willing to insure the utility companies owning the nuclear reactors for the amount that will surely need to be paid out in damages if an accident occurs.

It is important to recognize that this refusal to give proper insurance against a nuclear accident was based on what would happen to small 200 to 300 megawatt reactors situated thirty miles from the nearest city of 1,000,000. Yet in real life reactors are both much bigger than this and much closer to large urban centres. Most reactors are around 1000 megawatts in capacity and are being situated within 24 miles of New York City, 10 miles of Philadelphia and 4 miles of New London, Connecticut. In Britain, the Torness nuclear reactor complex is being built within 6 miles of the nearest town of Dunbar and is within 30 miles of Edinburgh.

Studies estimating the damage capability of these larger nuclear reactors have been made. The Committee for Nuclear Responsibility in California has done this, assuming a scenario of a 1000 megawatt reactor losing 10% of its radioactivity thirty miles away from the closest city of 1,000,000. The study also assumed that a substantial amount of the population would have been evacuated by the time the radiation cloud reached the city limits.[331] The findings include the following:

3,400 people would die immediately up to 100 miles away.

Assuming 600,000 other people received dosages of 100 rads, or 1,200,000 people received 50 rads, or 2,400,000 people received 25 rads, then 50,000 people would eventually die from the accident because of cancer or leukemia.

Birth defects, stillbirths, mental retardation could be expected

in higher than normal amounts in the children born to irradiated parents.
Agriculture and water supplies would be contaminated beyond use in an area larger than the state of California.

The AEC undertook a study of its own in 1964 to find out what an accident from a large reactor would entail.[332] Its findings indicate that a 'worst possible accident' would result in the following:

45,000 people would die.
100,000 people would be injured.
There would be $17,000,000,000 in property damage.
Poisonous fission products such as strontium-90 and cesium-137 would be deposited over a large area, contaminating air, water, soil, plants, animals and human beings. The study did not compute the long term consequence of this.

As in 1957, the AEC, upon finding out the extent of possible damage to environment and human life, reacted, not by evaluating whether the nuclear programme should continue in the light of this new data, but by suppressing the report. It was not until 1975, nearly ten years later, that the AEC was forced through the Freedom of Information Act to release the findings to the public.

Despite the potential devastation residing within the containment walls of every operating nuclear reactor, the atomic industry and the governments in virtually every country where the plants have been located have continued to assure the public that atomic power is so safe that 'no member of the public has ever died as a result of reactor operation'.

The industry and the government have sought to allay people's fears by saying that while a meltdown would indeed be troublesome, it will never happen. The reactors are simply built too safely for major accidents. Assuming perfection, studies such as the Rasmussen Reactor Safety Study have been highly publicized.[333] Conducted by physicist Norman Rasmussen of the Massachusetts Institute of Technology and costing over $3,000,000 over a two-year period, the study concluded that the chance of a single person dying from a nuclear accident is one in 5,000,000,000 per person per year – roughly 166,666 times less than any other accident conceivable. This means, the study said, that the chances of a person being killed due to nuclear power is the same as that of being killed by a falling meteor. This is to say, that nuclear power is so safe that there is nothing to worry about. Accidents will not happen.

Apart from the fact that this much-touted study used faulty computer methods, something even the US Nuclear Regulatory Commission (NRC) has finally admitted, what is important to be aware of is that nuclear reactors, like nuclear weapons, while indeed having a low probability of ever melting down or exploding, are nevertheless completely disastrous if they ever do. Furthermore, even given a one in 5,000,000,000 chance of a major accident occurring, this accident can just as easily happen today as in a million years.

One must finally, I think, agree with the statement made by the first Brookhaven study in 1957, that 'One fact must be stated at the outset: no one knows now or will ever know the exact magnitude of this low probability of a publicly hazardous reactor accident.'[334]

C. NUCLEAR ACCIDENTS

The first nuclear accident in the US was not only major but lethal. It occurred at the Stationary Low-Power Reactor No. 1 (SL-1) near Idaho Falls, Idaho, on 3 January 1961.[335] Only a 3 megawatt prototype military reactor, it had been shut down for work on instrumentation and the control-rod devices disconnected. Three young servicemen, John Byrnes, Richard McKinley and Richard Legg, were sent in to reassemble the control-rods. What this meant was that the central control rod was to be lifted ten centimetres and coupled to the remote driving mechanism, a simple procedure the men had done many times before. Yet on this particular day, things went wrong. Later reconstruction by the AEC indicates that the assemblage had taken place but that then – for reasons still unknown – the central control rod was pulled out of the core. AEC investigators suggest that perhaps it was stuck and Byrnes and Legg tried to pull it up manually. When it came loose, however, it rose not just ten centimetres but fifty, causing the entire reactor core to go supercritical. The fuel inside melted, causing a steam explosion that blasted a solid wall of water to the roof of the reactor vessel. The reactor vessel rose three feet, right through the pile cap. Legg and McKinley were killed instantly, McKinley's body being impaled on an ejected control rod plug on the ceiling. Byrnes was completely irradiated.

Simultaneously with the meltdown of the core and the deaths of the two men alarms rang. Emergency squads attempted to get near the reactor but while still some distance away their radiation dose meters were reading off-scale, meaning more than 500 roentgens per hour – a lethal dose. The level inside the reactor was even higher – 800 roentgens per hour. Despite this, two rescuers rushed into the wreckage and dragged out Byrnes. It was too late for him, however; he died en route to the hospital.

Recovery of the other two bodies had to be carried out with remote control handling gear, such was the level of radiation. Fourteen rescuers received more than their allowable five rems per year during the operation. All three bodies were so radioactive that it was necessary for three weeks to elapse before it was safe enough to handle them for burial. Even then, they had to be buried in lead-lined caskets and placed in lead-lined graves.

As for the reactor itself, the level of radioactivity was so high that investigators had to wait for several months before being able to go inside.

The nuclear reactor accident releasing the largest amount of radioactivity into the atmosphere occurred in Britain on 8 October 1957, at the Windscale No. 1 plutonium reactor.[336] On this occasion, a worker was carrying out a routine operation known as releasing Wigner energy, which raises and lowers the power level in the reactor core. According to his instruments, the core temperature was falling so he decided to give the temperature a boost. What he wasn't aware of, though as a worker in the area he should have been, was that the thermocouples recording the core temperatures were not in the hottest part of the core. When at 11.05 a.m. he raised the temperatures yet again, at least one fuel rod inside the core ignited. The worker went on unperturbed, however, and it was not until 5:40 a.m. on 10 October – some 42 hours and 35 minutes later – that there was any external sign that the reactor core was malfunctioning. At this point instruments began to indicate that radioactivity was reaching the air filters on top of the cooling-air discharge stack. These filters were known as 'Cockcroft's Folly', after Sir John Cockcroft, who had insisted they be installed – to the derision of his colleagues – as a precautionary measure. This

'Folly' kept the accident from becoming a major catastrophe, for by the time the Windscale staff realized anything was amiss, the reactor fire had become an inferno, spewing radioactivity everywhere. The filters on top of the stacks kept much of this contained.

Taken by surprise, no one was sure what to do. Molten uranium and cladding, full of radioactivity, was ablaze in over 150 fuel channels. Tom Tuohy, later the Windscale General Manager, remembers standing with a breathing respirator on and looking down through a viewing port above the cooling pool. What he saw were flames shooting out of the core and licking the concrete shielding of the outer walls – walls made of concrete which had specifications requiring that it be kept below a certain temperature lest it weaken and collapse.[337] At its height, there was eleven tonnes of uranium on fire.

Using water to extinguish the flames was out of the question because water and molten metal can react when put in contact – the metal oxidizing and leaving the hydrogen to mix with any incoming air and explode. Such an explosion could easily rip apart the containment walls. Another option was to try carbon dioxide, but at the temperatures the fire was raging it was possible that the oxygen of the carbon dioxide would feed the flames as effectively as air. This was tried, however, and sure enough, the flames only intensified.

By this time the fire had been raging out of control for over twenty-four hours. Early on the morning of Friday, 11 October, therefore, the decision was made to try water, despite the dangers of the hydrogen exploding. The Chief Constable of Cumberland was notified and warned of the possibility of an emergency. At 8:55 a.m. Tuohy ordered everyone out of the building except himself, one colleague and the local fire chief. He then turned on the water hoses.

It worked. Slowly the fire subsided and the task of estimating the damage began. The press and local townspeople had been told nothing until after a full day of the fire-fighting had elapsed. After the fire, however, it became apparent that thousands of curies of radioactivity had been released into the atmosphere. The questions came pouring in: how much radioactivity had been released over Westmorland and Cumberland? What kind was it, how dangerous? What could be done about it?

Many different types of radioactivity were released, the most lethal of which was iodine-131, which, as will be recalled, has a short half-life of eight days but is very active, causing cancer of the thyroid gland. It was estimated that 20,000 curies of this one type of radioactivity alone were released over an area of more than 500 square kilometres. Because iodine-131, like radioactivity in general, tends to concentrate in the food chain, cattle grazing in the contaminated area began producing irradiated milk. By arrangement between the British Atomic Energy Authority, the local police, the Milk Marketing Board and the Ministry of Agriculture, therefore, more than 2 million litres of milk were collected and dumped into the rivers and sea.

In the aftermath, an official inquiry was made, but the full report was never made public. What was made available was that design changes to prevent another fire were prohibitively expensive. The final decision was that both Windscale No. 1 and No. 2 should be permanently closed. Both were filled with concrete and entombed.

There have been other accidents in virtually every country with a nuclear programme. In the Soviet Union, for example, at the end of 1957 or the beginning of 1958, a disaster happened in the southern Urals involving a nuclear waste site. Although hardly anything is known about the details of the accident, according to Soviet physicist Zhores Medvedev, it was the result of a nuclear waste explosion caused by improper burial, not far from where the first Soviet military reactors had been built. Because there were strong winds at the time, radioactivity, particularly strontium-90 and cesium-137, contaminated an area of more than a thousand square miles. Entire villages were wiped out either through death or evacuation, and several hundred square miles are still uninhabitable. Villages within this restricted area have been completely bulldozed over to prevent the townspeople from attempting to return. One lake within the area is still contaminated with over 50,000,000 curies of radioactivity.[338]

In Canada, on 25 May 1958, an irradiated fuel element broke and caught fire inside a reactor near Chalk River, some 200 kilometres north of Ottawa. The resulting radiation release was estimated to be as high as 10,000 roentgens per hour, which necessitated over 600 workers coming in to help clean up. Around the plant, some 400,000 square metres were contaminated.[339]

In Switzerland, a similarly serious accident happened at the Lucens reactor on 21 January 1969, and was of such severity that investigators had to wait twenty-one months before the radioactivity levels lowered enough for them to gain access to the reactor core. Special equipment had to be designed and built to dismantle the reactor which was damaged beyond the hope of re-use.[340]

In West Germany, the first large commercial nuclear reactor was started on 2 October 1971, supplied its first electricity on 18 December 1971, and had its first major accident on 12 April 1972, when a pressure relief valve inside the reactor vessel got stuck in the open position – releasing over 200 tonnes of radioactive steam. It took seven months to clean up the contamination. The plant, built near Wuergassen, had to be shut down again on 25 February 1973, because major leaks developed in its primary piping. Undaunted, the company fixed things up, restarted operations in July, only to have new leaks develop. By February 1974, the plant had to be shut down again. It has not reopened.[341]

In Japan, perhaps the most bizarre and perversely humorous reactor accident publicly known occurred in 1974, with the nuclear powered cargo ship, the NS Mutsu. Powered by a pressurized water reactor, the Mutsu was launched in 1969 but it was prevented from actually even sailing out of its home port because of the intensity of public opposition. Most of the opposition came from local fisher-people who were deeply concerned lest the radioactive discharges of the ship contaminate the rich scallop fisheries of Mutsu Bay. Finally, in August 1974, the ship was allowed to leave its home port after the government agreed to set up a federal fund of 100 million yen to cover any compensation if the fisheries were contaminated. Many local officials and fisher-people remained adamant in their opposition, however, and as the ship set sail they blockaded it inside the harbour with over 250 small sailing craft. Because of bad weather conditions from an oncoming typhoon, the blockade did not hold, and the Mutsu eventually made it out to the open sea.

When the ship was 800 kilometres from the coast, the nuclear reactor was fired up. When the power was raised, however, leaks began to develop, but since the ship was being commissioned out at sea rather than in a port under the observation of the atomic energy authorities, the crew onboard had to improvise. Their first attempt to stop up the leak was to boil up some of their borated rice to be used as an impromptu shielding cement. When this failed, they gathered as many pairs of socks as they could find and used these to stop up the leaks. It would have obviously been more advantageous for all concerned if the ship had been allowed into a port for necessary repairs. But the opposition to even the ship's existence was so high across Japan that the captain did not dare to return to his home port. While the ship was drifting helplessly at sea, therefore, negotiations began to take place with local officials and fisher-people. It was finally agreed that the Mutsu could return home, but only on condition that it left permanently within six months and that another fund be set up to cover the damages resulting from its radioactive releases while in Mutsu Bay.[342]

Although the above account of nuclear accidents has selected one per country, it should not be assumed that this is all that has happened. Quite the contrary: accidents have been happening in all countries with nuclear reactors on a regular basis. In West Germany alone, six out of fourteen reactors are now shut down for either repairs or 'urgent inspection'. There have been over 140 nuclear accidents in Germany since 1961.[343] In Britain, radioactive releases from the Windscale reprocessing plant have been happening regularly, the latest one being the release of over 30,000 curies of radioactivity in thousands of gallons of liquid that were leaked into the ground. This occurred in April 1979. In the US the NRC asserts that there are over 200 'abnormal incidents' among the 72 operating reactors *per month*. Indeed, so frequent are nuclear reactor mishaps, leaks and accidents that the average reactor is shut down approximately 40% of the time. On 5 April 1978, for example, two technicians at the Trojan nuclear reactor near Prescott, Oregon, received at least 17.1 and 27.3 rems when a spent fuel assembly passed through the room they were in. This mishap was the *thirtieth* accident to have occurred at Trojan in a single six-month period.[344]

Such negligence is due not only to the poor design and maintenance of reactors but to the fact that the regulatory agencies supposed to be keeping a watchful eye on the atomic industry and the utility companies it serves do not so much regulate as *promote* the industry. All probabilities of accidents are minimized in order to assure that public nuclear power is safe. When accidents happen, therefore, the information is either suppressed or the damage done is disregarded as insignificant. The Swiss Association for Atomic Energy, after the January 1969 accident just described, not only attempted to distort information as to the extent of the actual damage but in its newsletter attempted to relegate the entire incident to insignificance by stating that, 'It had in any case been the intention to shut down the plant some time in 1970 or 1971. The incident can therefore be considered as an additional and very instructive experiment.'[345] When in the 30 June 1977 edition of *New Scientist* Zhores Medvedev revealed to the world the Soviet accident at the Kasli Atomic Plant in Zyshtym, in the southern Urals, an accident in which thousands of people died and were made homeless and hundreds of square kilometres became uninhabitable, the Chairman of the United Kingdom Atomic Energy Authority, Sir John Hill, attempted to dismiss the report as 'science fiction', 'rubbish' or a 'figment of the imagination' of Medvedev.[346]

The failure to effectively regulate in the US has become so serious that the following situation has developed. Every two and a half days in 1976, there was a reported failure of containment isolation at a nuclear reactor; a reported failure of main cooling occurred 116 times in that same year; failure of reactor protection 82 times; failures of other core and containment cooling systems 528 times. Altogether, there was one failure every seventeen hours, resulting in over 13,000 workers being exposed that year in excess of 500 millirems.[347] Yet the utility companies do not feel pressured by anyone to improve their performance. The government agency mandated to regulate and inspect them simply does not do its job. On 30 June 1974, for example, the AEC reported that for the previous year it had discovered 3,333 safety violations at the 1,288 nuclear facilities operating in the US. 98% of these violations posed a threat of radiation to the workers. Despite this fact, the AEC only imposed punishments for *eight* of the 3,333 violations.[348].

The corruption of the AEC became such a scandal that in 1975 the AEC was dissolved by an act of Congress. In its place the Nuclear Regulatory Commission was set up. Yet nothing changed, perhaps because 87% of the top decision-making officials of the NRC are former nuclear industry officials.

It is within this context, then, of accidents happening on a regular basis but with no governmental agency effectively attempting to ensure the safety of the plants, that the nuclear industry is allowed to exist and grow. Because of this negligence, the probability of the 'worst accident possible', a meltdown, has increased far beyond the one in 5,000,000,000, projected by the Rasmussen study. Indeed, meltdowns have already occurred some of which have been described. The one drawing the most public attention, however, perhaps because it most directly threatened thousands of citizens, has been the recent 'incidence' at Harrisburg, Pennsylvania.

D. HARRISBURG

The Three Mile Island 960 megawatt pressurized water reactor was as plagued with problems as most nuclear reactors are. During 1978, it was closed down for repairs for 195 out of the last 276 days of the year. Nevertheless, Metropolitan Edison (Met-Ed), the utility company owning the plant, still managed to get the plant 'on line', meaning generating electricity to consumers, by 30 December 1978, the last possible day to qualify for a $37-48 million tax-investment credit from the federal government. The design of the Three Mile Island plant can be seen in the following diagram:[349]

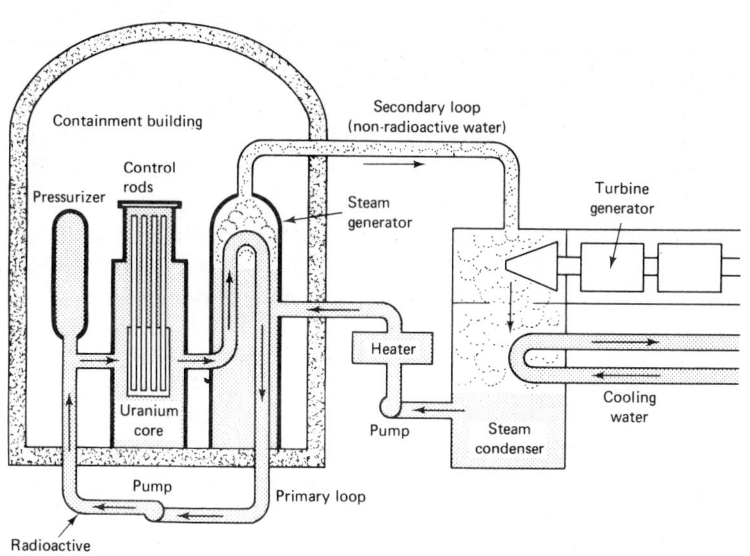

Design of Three Mile Island Reactor

This diagram indicates the closeness of the reactor to neighbouring communities:[350].

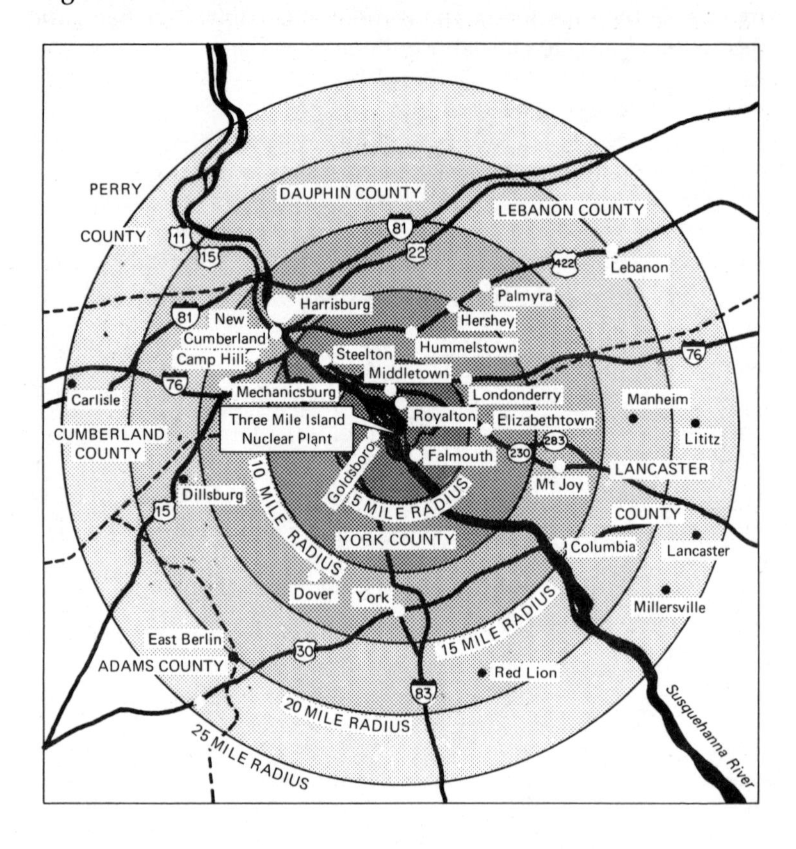

Two weeks after the plant had been on line, in mid-January, two safety valves ruptured, closing the reactor down for two weeks. Then on 1 February a throttle valve developed a leak; on 2 February a heater pump blew a seal; on 6 February, a pump on a feed-water line tripped off. Other similar types of accidents and mishaps followed. During maintenance operations in mid-March 1979, valves on three or four of the back-up pumps were left closed, in specific violation of NRC regulations that because they are critical for any emergency shut-down they must always be left opened while the reactor is running. Yet for over two weeks no one even noticed the violation. When it was noticed, it was too late.

On Wednesday 28 March, the reactor was running at full capacity. The regular four-person crew sat in the control room keeping watch over a forty-foot panel along three walls lined with gauges, dials and over 1,200 warning lights colour-coded red and green. Everything seemed to be operating smoothly until shortly before 4:00 a.m. According to the NRC later, what occurred at that point was somebody 'screwing around with some of the equipment in the feedwater system; whatever he was doing resulted in tripping the feedwater pumps off the line'.

The feedwater pumps send hot water to the steam generator as well as send cooling water into the reactor. With both pumps shut down, a sensor in the steam generator realized that it was no longer receiving water and so it immediately shut down as well. This in turn shut down the plant's turbines. The turbines shut down, the reactor electronically sensed that it did not want any more steam, and so a switch was automatically turned on which shot a jet of steam up from the plant's turbine building at a pressure of 1,000 pounds per square inch. From the turn-off of the feedwater pumps to the jet of steam took only six seconds.

At this point a relief valve automatically opened to blow off extremely hot radiated water inside the containment vessel. Altogether twelve seconds had elapsed when the reactor 'scrammed' the control rods inside the core to stop the chain reaction process. This immediately slowed the fissioning of the uranium-235, in effect shutting down the reactor.

The pressure inside the reactor vessel began to fall, which should have signalled the relief valve to close. But it did not, allowing the pressurized steam supposed to cool the core to pour *out* of the core area into the surrounding containment area.

The main pumps having failed, three auxiliary pumps were immediately turned on. But the valves that should have opened to allow their water to come into the reactor core failed to open. It was these pumps that had been closed some weeks before during 'routine maintenance', in violation of one of the NRC's most fundamental rules that they should *never* be closed while the reactor is operating, just in case they are needed for any emergency.

Without enough water, temperatures inside the core began to rise, climbing thirty degrees in less than three seconds. This caused more of the water inside the core to vapourize and shoot out of the still-opened relief valve.

All of the above had transpired within two minutes.

When the water pressure inside the reactor had fallen to 1,600 pounds per square inch (psi), the plant's emergency core cooling system was automatically turned on. For some reason, however, an operator in the by-now panicking control room *shut off* the system's two pumps. He shut one down four minutes thirty seconds into the accident, the other one six minutes thirty seconds into the accident. The reason for this action remains a mystery. It has been suggested by both the NRC and Babcock and Wilcox, the designers of the reactor, that the operator was only looking at one of the two gauges he should have checked.

This left the reactor core without any incoming water, only vapourized steam still spewing out of the relief valve still stuck open. Only seven and a half minutes into the accident the highly radioactive water spurting out of the reactor core was two feet deep in the containment area.

This flooding automatically triggered the building's sump pumps which began pumping water out of the containment building into tanks in an auxiliary building.

Eight minutes into the accident an operator realized that the valves allowing the auxiliary pumps to drive water into the core were off. He quickly turned them on, thus allowing water into the core area for the first time. Three minutes later, it was realized that the Emergency Core Cooling System was out as well. This was now turned on, and for the next fifty minutes water came pouring into the reactor core, stopping the fall of the reactor pressure and lowering the temperature. Thus far there was still enough water to cover the 36,000 fuel rods that make up the core itself, meaning that things were still basically under control.

But then the inexplicable happened again. An operator turned off four cooling pumps, two one hour fifteen minutes into the accident, two one hour forty-five minutes into the accident. This time the mistake was fatal. The water level and pressure dropped so far that a third of the core became exposed. Fourteen minutes later, the temperatures of the exposed fuel rods climbed right off the scale. In the control room the computer monitoring the core temperatures printed readings of 750 degrees centigrade and then began printing question marks. The meltdown had begun.

The stainless steel coating around the uranium pellets, called cladding, began to crumble and highly radioactive particles and rays began to pour out. It is estimated that 20,000 rods were oxidized, radiating the water still pouring out of the still-open relief valve.

Shortly before 7:00 a.m., the emergency siren at the plant began to sound, signalling workers to evacuate certain critical areas. A number of workers attempted to drive their cars away from the plant entirely. Two made it out before the gates were locked. At 7:02 a.m., company officials notified the civil defence officials of Dauphin County that they had declared a plant emergency. Back in the control room, however, the radiation meters indicated that radioactivity inside the containment area had exceeded eight rems. This meant more than a plant crisis; radiation levels this high signified a general emergency of all of Three Mile Island.

The necessary local officials were notified of this at 7:20 a.m. At 10:00 a.m. Dr Harold Denton, the chief of reactor operation for the NRC, was notified that a 'relatively serious sort of event' was happening. At about the same time also, the White House was notified, the news being relayed to Zbigniew Brzezinski, National Security Advisor to the President. He took the news immediately in to President Carter.

At this point, when the danger was reaching the critical stage and the lines of command and evacuation procedures should have been the clearest, 'there was just a terrible communications problem,' Dr Denton was to recall. 'All the phone lines were jammed up. . . . You got only bits and pieces.'[351]

Despite this, federal and NRC officials remained fairly unperturbed, feeling that the situation was basically in hand. 'We had a rough sequence of things that had gone wrong, we thought,' said Denton. 'We didn't know what the cause was. I thought it had been a small loss-of-coolant accident.'[352]

Inside the control room, things were different. By mid-morning, with heat readings in the core going off the scale, the computer still printing out question marks and the level of radioactivity rising, officials were getting desperate. At 11:30 a.m. they decided to 'blow down' the system, which involved trying to reduce the pressure of the cooling system to 400 psi. This level would allow the operators to turn on the large water pumps normally used to bring a reactor to 'cold shutdown'.

At first all went well, but at 2:00 p.m. a hydrogen explosion jolted the reactor, setting off the emergency sprays near the ceiling of the containment structure. They began pouring 5,000

WHAT WENT WRONG ON THREE MILE ISLAND

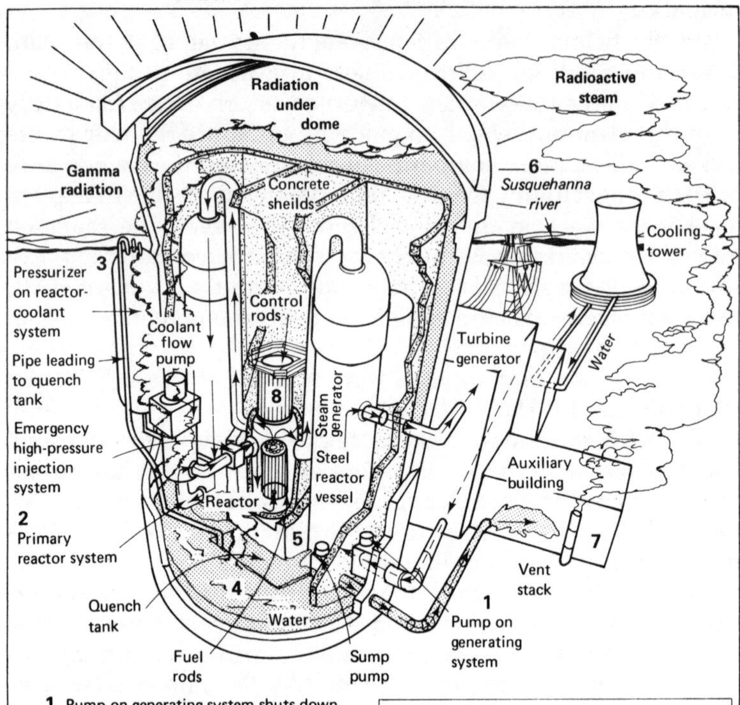

1 Pump on generating system shuts down, cutting off water to steam generator.

2 Reactor keeps producing heat, raising the temperature and pressure of the water in the primary reactor system. The reactor shuts down automatically.

3 Valve on pressurizer opens as planned, but fails to close. Radioactive water gushes into quench tank, which overflows and floods the floor of the containment structure. Some radiation penetrates wall of the structure.

4 The pressure and water level in the reactor system drop, triggering the emergency cooling system. But an operator shuts it off. Some fuel rods overheat and burst–or perhaps even melt.

——→	- - - - - - →
Reactor-coolant flow in primary system	Steam and water flow in generating system

5 Sump pump transfers radioactive water to nearby auxiliary building. The building floods, and radioactive steam is vented.

6 The next day, radioactive water is dumped into the river.

7 More radioactive material is released from auxiliary building.

8 A radioactive gas bubble forms at the top of the reactor. It raises the danger of an explosion or fuel-rod meltdown unless it can be vented within a few days.

gallons of white sodium hydroxide all over the reactor. What the hydrogen explosion had done was what the officials had feared: a bubble of explosive gas had formed inside the reactor. This new situation caused them to stop attempting to de-pressurize the system.

At 5:30 p.m., the Met-Ed officials decided to bring the pressure back up in an effort to burst the bubble. They also decided to try to restart the main reactor coolant pump which had shut down when the accident began. When this pump started up, water began pouring into the reactor, covering the fuel rods, which by now had been left exposed and disintegrating for more than eleven hours. The temperatures began to drop and the pressure held at around 1,000 degrees Centigrade.

The extent of the damage to the reactor, however, had by now become quite extensive, as the diagram on page 156 illustrates.[353]

It should be observed that at this point none of the radioactive releases from the reactor vessel had escaped the containment building, except the water being pumped into an auxiliary building.

On Thursday morning, 29 March, officials of Met-Ed, confident that the situation was under control, began a public relations campaign to assuage a confused and frightened public, particularly those living immediately around the plant. The mayor of the nearby town of Middleton, however, a Robert Reid, was unmoved by the company's reassurances and demanded to know why he had not been notified until after three hours into the accident.

Meanwhile, radioactive gas and steam were building up to critical levels in the auxiliary building where the sump pumps were pouring the radioactive water from the containment building. To solve this problem, officials opened vents and allowed the radioactivity to shoot out into the atmosphere, potentially contaminating a public that they were attempting to reassure that everything was under control.

Later that same day, Dr Ernest Sternglass, who, it will be recalled, conducted several of the studies of the impact of fallout during the 1960s, came down to the area on his own initiative to take samples at the Harrisburg airport, three miles north of the plant. The samples showed radioactivity levels *fifteen times* the normal amount expected. Dr George Wald, a former Harvard biologist and Nobel laureate, warned that radioactivity has adverse effects on pregnant women and children. This began calls from all over the region by those who felt they could have been damaged. Besides releasing radioactivity into the atmosphere, however, the plant officials had also done something that Dr Sternglass did not check: they began dumping thousands of gallons of radiated water into the Susquehanna river; this, because their holding tanks had become 'too full' from all the radiated water they were pumping out of the containment building. No one bothered to tell the communities downstream who drink this water.

Early the next day, Friday 30 March, a small monitoring airplane circling the plant picked up the fact that fresh radiation was coming out of the stack alongside the auxiliary building. The level was put at 1,200 millirems. This, too, was a deliberate venting of radioactive gas into the atmosphere by company officials to relieve pressures inside the building's holding tank. However, the company notified neither the state nor the federal officials nor the NRC about what they were doing, thus setting in motion a frantic effort on the part of state and city officials to protect the residents of the area.

Around 9:00 a.m. Harrisburg radios informed listeners there was an 'uncontrollable release of radioactivity' coming from the plant. At 10:00 a.m., Governor Thornburgh urged everyone within a ten-mile radius of the plant to stay indoors until further notice. The civil defence sent loudspeaker trucks throughout the area to warn residents. Schools were called, recesses cancelled, and no one was allowed out of the buildings. At 11:15, air raid sirens began to wail across the city of Harrisburg. There was some hysteria and panic in the streets, although most people remained composed. At the same time, the phone system in nearby Middleton, overloaded for some time, went dead.

Meanwhile John Herbein of Met-Ed began his long delayed press conference by saying that 'conditions are stable', explaining that the company had deliberately vented the radioactive gas into the atmosphere from 7:30 to 8:15 that morning. He added: 'It's certainly the civil defence's prerogative to take those steps (referring to asking people to stay indoors), but we don't think it was necessary. If the civil defence chooses to tell inhabitants of Middletown to keep their windows and doors shut, that's their prerogative. We have our doors and windows wide open.'[354]

Almost lost in his remarks was any mention of the hydrogen

bubble which was slowly building up inside the reactor vessel. 'It's serious,' he said, 'but not to the extent we have to evacuate the citizenry.'[355]

The extent of the radioactive gas release can be seen from the following diagram:[356]

THE DANGER ZONE

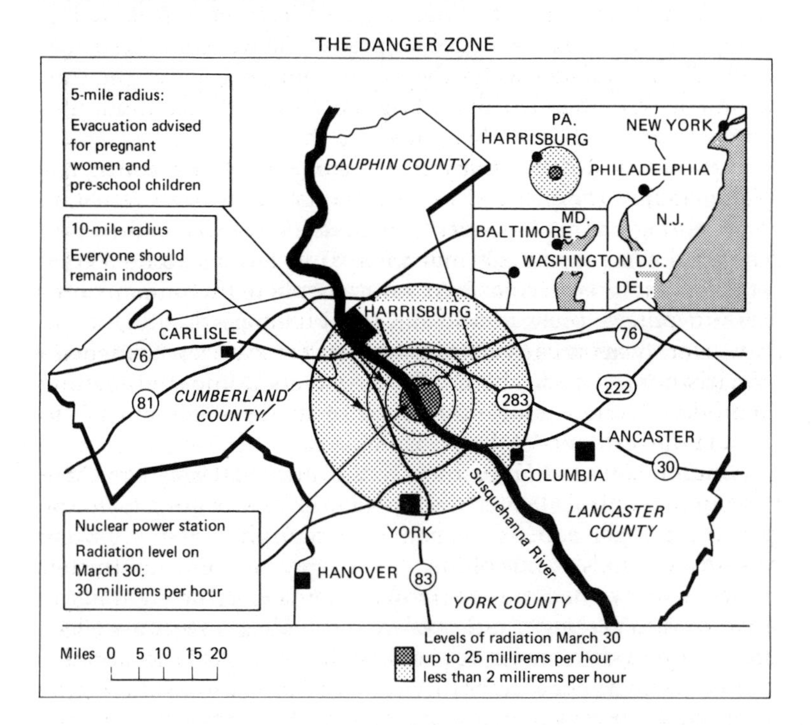

In Washington, NRC Chairman Hendrie was in confusion and near-panic. According to official transcripts, his initial reaction to the news about the radiation release was the following: 'We are operating almost totally in the blind, his (Governor Thornburgh's) information is ambiguous, mine is non-existent – I don't know, it's like a couple of blind men staggering around making decisions.'

Although describing the radiation released as 'a pretty husky dose rate', Hendrie could not make up his mind as to whether an evacuation order was appropriate or not. In Harrisburg, however, Governor Thornburgh came to a decision. He was advised by the Pennsylvania Department of Environmental

Resources to take the situation seriously and order a partial evacuation. After conferring with President Carter, the Governor called reporters to the state Capitol building and announced:

> I am advising those who may be particularly susceptible to the effects of radiation, that is, pregnant women and pre-school-age children, to leave the area within a 5-mile radius of the Three Mile Island facility until further notice.[357]

Meanwhile, the bubble continued to grow and by Friday night, 30 March, the possibility of a complete meltdown had become seriously real.

At this point, federal and state officials began taking over control of the situation. Presidential Assistant Jack Watson authorized the Food and Drug Administration to contract for the manufacture, packaging and shipment to Harrisburg of 240,000 one ounce vials of potassium iodine, a drug that can be ingested to saturate the thyroid gland against iodine-131 poisoning. Watson also directed Robert Adamcek, a regional director of the Federal Disaster Assistance Administration in Philadelphia, to co-ordinate evacuation plans with Governor Thornburgh. John McConnell, Associate Director of the Defence Civil Preparedness Agency, was also sent to consult with county officials about evacuation procedures. Blankets, cots and dosimeters (devices to measure radiation exposure) were gathered and sent to evacuation centres, and seventy tons of lead bricks were sent from around the country to the plant itself, should the situation demand emergency construction of radiation shielding. The price tag for these minimal preparations: $1.7 million.

The concern about a possible complete meltdown increased when it was discovered that fully one-third of the core was still exposed, causing the temperatures to go so high that the coolant water was decomposing into its primary elements: oxygen and hydrogen. This decomposition fed the ever-enlarging hydrogen bubble at the top of the reactor vessel. The biggest danger was that the bubble would grow until all the water was forced out of the reactor, allowing the temperatures of the fuel rods to build up to 5,000 degrees. At this temperature, the uranium would begin to melt – beginning the 'worst possible accident' described earlier – a meltdown. At 5,000 degrees, the heavy metals such as uranium and strontium, having melted together, would begin to burn their way right through the floor of the reactor vessel. The rest of the core, when it melted, would fall into a four foot deep pool of water that would still have remained below the fuel rods at the bottom of the reactor. The impact of a melting core hitting water would cause a steam explosion that could easily blow the reactor ceiling off like a missile, breaking open the containment walls outside. This would allow the radioactivity to be released directly into the atmosphere. It was estimated by officials that there would have to be at least 60,000 thyroid gland operations to cure the radioactive after-effects of the iodine-131 alone; hence, the ordering of the 240,000 one-ounce capsules of potassium iodine by Presidential Assistant Watson.

As it was, the 20,000 exposed fuel rods were still disintegrating, releasing massive amounts of radioactive gases like xenon-133, krypton-85 and iodine-131. Herein lay the second fear: that the hydrogen bubble would explode with enough force to shatter the structure around the reactor core, thus allowing these gases into the containment area. The NRC computed that a bubble as big as this one was – 1,000 cubic feet – could set off an explosion equal to three tons of TNT, which meant that it could conceivably not only blow open the reactor vessel but break open the four-foot-thick concrete walls of the containment structure as well, releasing radioactivity directly into the air. Such an event would easily cause a total loss of coolant around the core and trigger the meltdown they had been assuring the public for the past twenty years was too improbable ever to worry about.

At the height of the crisis on Friday, with no one sure whether a meltdown was going to happen or not, the NRC took over and gave Dr Denton full responsibility and control. The Met-Ed officials, however, still trying to minimize the seriousness of the situation, came into open conflict with him, stating to the press that the bubble had in fact shrunk from 1,000 to 800 cubic feet. Met-Ed spokesperson Herbein stated: 'Personally, I think the crisis is over.'[358] When Denton had *his* press conference, however, he stated quite categorically that the bubble had *not* shrunk and that the crisis would not be over until there was a complete state of cold shutdown.

Meanwhile in Washington, NRC Chairperson Hendrie issued a warning that residents within twenty miles of the plant should be prepared to evacuate. This stretched the perimeter for evacuation to over 630,000 people.

The final straw came just before 8:30 p.m. on Friday night. The Associated Press released a story stating that the bubble situation had become 'potentially explosive', citing experts who warned that it could explode at any moment. The reaction to this news can be exemplified by what Paul Critchow, Governor Thornburgh's press secretary, remembers of the reporters waiting outside his office. When the AP story came over the wires, he said,

> about 20 or 30 reporters burst through the door of this office. They said: 'We want to know if our lives are in danger. What the hell's going on here? We want to know if we have to get out'. . . . They were pale. They were frightened. At that point, they had lost all interest in the story they were supposed to be covering.[359]

The bubble did not explode. Gradually, ever so gradually, and of its own accord, it began to recede. By Sunday, it had shrunk to 650 cubic feet, at which point President Carter came to the plant to tour and inspect the damage. Over the weekend, however, thousands of people, variously estimated from 80,000 to 200,000 were voluntarily evacuated to safer areas.

On Monday morning, 2 April, the bubble had nearly dissolved, and the temperatures inside the reactor were receding. By that Friday, 6 April the state Office of Civil Defence estimated that 90% of those who had evacuated the area had returned home. Pregnant women and pre-school age children, however, were still being advised to stay away until further notice.

The after-effects of the accident seemed to add insult to injury. Evacuees found out upon returning home that they would have to pay for the accident that had victimized them through various rate increases in their monthly utility bills. The average increase was to be about $13.50 per customer per month. Moreover, the rate payers were notified that they would face additional rate increases to help pay for the clean-up of the plant.

According to experts in the NRC, the loss of the reactor core alone came to between $50 and $100 million. Cleaning up all the radioactive contamination, which means stripping everything containing any radioactivity at all completely away and burying it at one of the three radioactive waste disposal sites around the US will cost additional tens of millions of dollars – in fact, so much additional money that Met-Ed has decided not to attempt a clean-up at all. This means that the entire reactor complex must be disposed of.

This can be done in three ways. Met-Ed can attempt to dismantle and destroy the reactor itself, taking it apart piece by piece and shipping the pieces to burial sites. Another option is 'mothballing' the reactor – pulling out the usable parts like the turbine generator and then leaving the rest for some future generation to take care of. (Ten major US nuclear reactors have been mothballed.) The third alternative is 'entombment', which means sealing the entire complex underneath a concrete shell. (This was done to the Windscale plants after the 1957 fire: three US plants have met a similar fate.) As one nuclear clean-up specialist from the NRC describes entombment: 'It's a little like King Tut's tomb. You pour concrete over the entire thing, walk away and leave it sealed up for a thousand years.'[360] Hopefully by that time most of the radioactivity will have dissipated away. Most of it, that is, except for plutonium: plutonium remains radioactive and extremely dangerous for *250,000,000 years*. The NCR estimates at least 50,000 man rems of radiation exposure to the workers involved in the disposal of TMI. According to Dr Karl Morgan, this will result in approximately 30 additional cancer deaths among these workers.

The price tag for such procedures of nuclear reactor disposal vary from $42 million to the cost of what it took to build the plant in the first place, generally around $2,000,000,000 to $4,000,000,000.

There has been other economic fallout as well. Real estate prices have plummeted; tourism has dropped; and food markets have begun posting signs saying 'We don't sell Pennsylvania milk'.

For many people, the situation had been so traumatizing that they sought out professional help. Psychologists reported being inundated with calls. There was for most a wrenching realization that what they had been encouraged to trust as safe by their utility company, their local, state and federal governments, by the nuclear industry and by the federal regulatory agencies involved, was in fact *not* safe. People felt betrayed. Before the accident, the population around the Three Mile Island plant had supported the reactor being there, believing what they were told about it – that it was safe and would increase employment and local business. But the accident changed all this. As a front page editorial in the *Middletown Press and Journal* declared: 'If our faith in Met-Ed is shaken, our belief in the entire nuclear power industry rides on thin ice too.'[361]

Perhaps the most troubling aspect of the Harrisburg accident, however, has been the growing realization on the part of local residents that the radioactivity they were exposed to during the accident, particularly by the deliberate releases caused by Met-Ed officials, is going to have long-term effects upon them. They are entering, therefore, the same psychological space as the survivors of Hiroshima and Nagasaki were forced into – an awareness that they have been thrust into a situation of death-dominated life, for the radioactivity they were exposed to during late March 1979, will only begin to reap its effects in the years to come with cancer, with leukemia, and with a higher-than-normal rate of genetic defects in their children. This feeling is probably summed up most succinctly in a slogan that began appearing on T-shirts and bumper stickers shortly after the accident: It read: 'I survived Three Mile Island . . . I think.'

More than a year after the accident, however, most residents are realizing that they in fact did *not* survive the accident without damage and traumatization to their lives. There have been dramatic increases in thyroid cancers and stillbirths; farmers have complained of mutated livestock and contaminated well water; and there have been reports of increases in all sorts of psychological disorders, most of which relate to the stress people have felt since the accident.

On 11-14 April 1980, Citizens' Hearings for Radiation Victims were held in Washington, DC to bring to public attention the fact that the nuclear fuel cycle from uranium mining to nuclear weapons testing has and is producing massive numbers of victims from radiation exposure. Among those testifying was Susan Shetrom, who lives within 3 miles of the plant and has therefore experienced the full impact of the Three Mile Island accident. Her testimony was both moving and representative of the feelings of many Pennsylvania residents in the aftermath of an accident the government and the nuclear industry continue to insist was our 'safest accident':

> The accident at Three Mile Island is not over. It will not be over for us as long as the clean-up of Unit 2 continues. Nor will it ever be over if Unit 1 is permitted to restart. Because the greatest obvious damage thus far has been psychological, we have lived with fear for more than a year. We will con-

tinue to be afraid for the rest of our lives. We can see the source of our fears every day. Since the accident there have been numerous releases of radioactive gases into the air and of radioactive water into the Susquehanna River. And we are painfully aware that the damaged reactor is still capable of another, possibly more serious, accident.

I live with my husband and seven year-old daughter within sight of Three Mile Island – just three miles away. I have lived there for thirteen years – long enough to remember the construction of the plant – long enough to remember the promises of safe, cheap energy. Instead, we have become victims; physically, psychologically, and financially, of the nuclear industry's madness and greed. In 1974, when Unit 1 began operating, I was a typical conservative York Countian who trusted the government to act responsibly. I had never protested, attended a rally, or admonished an NRC commissioner. Until 28 March 1979, I continued to trust.

At that time I was not aware that the number and placement of dosimeters were inadequate, that eight environmental samplers had not been calibrated since 1974, or that for the two days of March 28th and March 29th the radiation monitors were all off scale. I am now convinced that those releases of radioactivity have doubled my family's chances of getting cancer, and that the chances of my daughter someday bearing a healthy child have been greatly diminished. I no longer tuck her into bed with lovely thoughts of her future, but rather with the fear of the results of radiation damage that will make themselves known in the years to come.

Already there have been reports of thyroid disorders and a rising infant mortality rate. Many area farmers have seen their animals die of diseases attributed to the effects of radiation, not just since the accident, but ever since the plant began operating. Our state health department does not yet agree that there is any connection between these and the accident. But the whole emphasis both locally and nationally has been to downplay the health concerns. Also, it is a fact that it may take from 15 to 30 years before cancer appears after the cell is exposed to radiation. We will not know the full results of the on-going accident for many years. So, to make statements, as Metropolitan Edison has, that no one was hurt, is absolutely ludicrous.

In the meantime, I have been doing what I can to see that

neither reactor ever operates again. This has meant the sacrifice of many precious hours I could have spent with my family. I am fortunate (although some of my friends are not) in that my husband supports my efforts. But how do you explain to a seven year old why mommy isn't home much anymore? Why there is no time for the activities they used to enjoy together? Why there are tears in her mother's eyes when they hug? I am keeping a diary for my daughter so that when she's older she'll understand why I couldn't sit idly by while the nuclear industry, with the full support of our government, commits random murder.

By being outspoken, I've also sacrificed some friends. There are many diverse opinions in the Harrisburg area – and, therefore, there is much hostility. Some people refuse to believe that there was or is an accident; they block the whole incident out of their minds. To them, I am an unpleasant reminder, easier to harass than believe. Some say they believe in fate – there's nothing we can do – you won't die until God wills it – what will be, will be. I believe in a God that expects us to use our intelligence. What I don't believe in are the Nuclear Regulatory Commission and Metropolitan Edison. They have insulted our intelligence, laughed at us, sneered at us, and lied to us. Since the accident there have been several so-called 'accidental' releases of water and gases. Each release is more difficult to cope with. I even carry a metal box containing our valuable papers with me in case something happens while I am at work. Through it all we hear helicopters and sirens. Each time a siren begins its wail I hold my breath and count, wondering if we'll have to evacuate again. I understand there is now an evacuation plan (only a year and two weeks too late). According to some authorities it is better than what we are told we had, but not necessarily workable.

I know people who have great, almost uncontrollable anger. Some people have sought psychiatric help. And still others have required medical attention because of the physical effects of trauma. Our children are unhappy and afraid. They have nightmares, cry frequently, are irritable, and want to be held constantly. Their sense of security has been shattered. And recent studies by an independent researcher have shown that stress levels are continuing to rise.

Depending upon my activities, comments by others, and

conditions at the plant, I find that I am either very high, very irritable, or very depressed. Along with these go any combination of tense muscles, stomach cramps, insomnia, twitching eyes, nightmares, and forgetfulness. I've actually found myself standing in the shower for as long as twenty minutes just wringing my hands.

Perhaps we should move. Every week I learn of more families who have. And who have had significant improvements physically and mentally. For my family moving would mean literally sacrificing a home, two jobs, and many dreams. Yet I have pleaded with my husband to do just that. And then we sit down and discuss where to go. We haven't found a place far enough from another reactor, a uranium mine, or a waste disposal site. And we also know that moving wouldn't undo the damage already done. So we stay – and fight – and wait. And waiting for the results of the accident at Three Mile Island is a twenty-four hours a day nightmare.

E. THE EFFECTS OF LOW-LEVEL IONIZING RADIATION

This gnawing concern that the impact of the Harrisburg accident has produced the same 'invisible contamination' as that produced by the atomic bomb over Hiroshima and will in the years to come strike down its victims with cancer, leukemia, and genetic defects is something that needs to be examined, particularly in the light of the fact that Met-Ed officials felt free deliberately to dump radioactivity into both the atmosphere and into the Susquehanna river. In order to understand, therefore, why on the one hand the population surrounding the plant now feels a long-term threat from those releases and why, on the other hand, Met-Ed felt no compunction at all in allowing the releases, it is necessary to know one final and very important fact about radiation.

For many years, the majority of scientists accepted the theory that despite the damaging impact of ionizing radiation upon living cells, there was nevertheless a *threshold* level in the body under which there was no danger of damage. This meant that there was in fact a safe level of exposure to ionizing radiation. If a person did not exceed this threshold no harm would result, as the radiation damage would be healed by the body's healing mechanisms. The person most responsible for articulating this threshold theory in the last several decades has been Dr Karl Morgan, called the 'father of health physics' because he was so instrumental in setting the existing standards for radiation exposure – 5 rems per worker per year and 0.17 rems per member of the public per year. He was the Director of Health Physics Division at the Oak Ridge National Laboratory from 1943-1972 and has also served as Chairperson of the Internal Dose Committee for the International Commission for Radiation Protection (ICRP) and the National Council on Radiation Protection.

All nuclear reactors, indeed, all nuclear facilities of any type, were built upon the basis of the threshold theory. Accepting this theory, that there are allowable levels of radiation exposure, Met-Ed felt it could dump the radioactivity it did into the river and the air around the Harrisburg reactor.

In an article published in September 1978, however, entitled 'Cancer and Low Level Ionizing Radiation', Dr Morgan asserted that

> From 1960 to the present, an overwhelming amount of data have been accumulated that show there is *no safe level* of exposure and there is no dose of radiation so low that the risk of a malignancy is zero. . . . For man there is never a complete repair of the radiation damage, since even at very low exposure levels there are many thousands of interactions of the radiation with cells of the human body.[362]

To illustrate the point, Dr Morgan points out that even a normal diagnostic X-ray of one rad emits 2,200,000,000 photons per square centimetre to the part of the body exposed. 'It is inconceivable', he says, 'that all the billions of irradiated and damaged cells would be repaired completely or replaced.'

Dr Morgan goes on to say that 'the most significant damage from low-level exposure results from direct interaction of the stream of ions with the nucleus of one of the billions of irradiated cells that may . . . survive and continue to divide but fail to repair the radiation damage'. It is in each nucleus that the 46 chromosomes lie, each one of which is coded with millions of bits of genetic information. If mutated, Morgan says, what can occur is that the 'cell survives and multiplies in its perturbative (mutated) form over a period of years (5 to 70 years) and forms a clone of cells that eventually is diagnosed as a malignancy'. This malignancy can turn out to be leukemia, a variety of cancers, or, if the chromosome damage is in the reproductive organs of the body, in birth defects.

What this means, according to another physicist, Dr Walter Patterson, is that 'the danger of radiation to living matter seems to increase in direct proportion to the amount of radiation exposure, beginning from the very lowest dosages'.[363]

This information is not something new but has been building up over the years. Indeed (as the table below indicates) the

Changes in Levels of Permissible Exposure to Ionizing Radiation

For Radiation Workers

Recommended Values		*Comments*
0.1 erythema dose/y (–1R/wk for 200 kV X-ray	52 R/y	1925: Recommended by A. Mutscheller and R. M. Sievant 1934: Recommended by ICRP and used worldwide until 1950
0.1 R/day (or 0.5 R/wk)	36 R/y	1934: Recommended by NCRP.
0.3 rem/wk	15 rem/y	1949: Recommended by NCRP. 1950: Recommended by ICRP for total body exposure
5 rem/y	5 rem/y	1956: Recommended by ICRP 1957: Recommended by NCRP for total body exposure

For Members of the Public

Recommended Values		*Comments*
0.03 rem/wk	1.5 rem/y	1952: Suggested by NCRP for any body organ
0.5 rem/y	0.5 rem/y	1958: Suggested by NCRP 1959: Suggested by ICRP for gonads or total body
5 rem/30y	0.17 rem/y	1958: Suggested by ICRP for gonads or total body
25 mrem/y	0.025 rem/y	1977: Suggested by EPA [20] for any body organ except thyroid*
5 mrem/y	0.005 rem/y	1974: Suggested by ERDA for persons living near a nuclear power plant **

R = roentgen. 1 R = 0.88 rem
rem = roentgen equivalent man
mrem = millirem
NCRP = National Council on Radiation Protection and Measurements
ICRP = International Commission on Radiological Protection

*The limit set by the Environmental Protection Agency for the thyroid was 0.075 rem per year.
**Present radiation protection guide of the Nuclear Regulatory Commission.

scientific community has been forced over the decades to continue to lower the allowable dosages of radiation, the more the scientists discovered about it.[364]

The newest evidence indicates, as both Morgan and Patterson point out, that *any* radiation affects the body, even the minutest of neutrons, alpha or beta particles, or gamma or X-rays. *There is no threshold under which a person can believe he or she will remain safe from damage*. Hence the fear of the residents around the Harrisburg plant who breathed in the radioactive air and drank the radioactive water.

The contradiction, however, is that nuclear reactors cannot operate without the threshold theory, for radioactive releases are inevitable, occurring even during normal day-to-day operations. This means that *for nuclear reactors to continue operating, people will have to be prepared to suffer increases in cancer, leukemia and birth defects rates*. This holds true for the workers working in the plants and for the public living around the plants.

What is important to point out, however, is that the radiation released by nuclear reactors and the corollary damage to human and environmental life is only one aspect of the entire sequence. At each stage in the progression of the uranium, from the time it leaves the ground till it returns to the ground to be buried as waste, there are radiation releases and therefore dangers to the public and the environment.

F. THE NUCLEAR FUEL CYCLE

1. Mining

The uranium pellets that go inside the 36,000 fuel rods that make up the reactor core of a nuclear power plant have gone through a long process of chemical treatment and enrichment before they arrive at the reactor. The process begins with the mining of uranium ore. In the mining of uranium, a gas, radon-222, is produced. An alpha emitter, radon-222, produces polonium-218 as a 'daughter product', a particle which clings to the dust particles inside the mine breathed in by the miners. If ingested, these radioactive agents can cause various types of cancer, most commonly small-cell and/or pulmonary carcinoma.

High cancer death rates among uranium miners has long been known about. Miners in Europe's earliest pitchblade mines of

Joachimsthal and Schneeberg were known to suffer from a mysterious disease they named 'mountain fever'. More than half the miners died of this 'fever', later identified as lung cancer. Between 1875 and 1912, of the 469 reported miner deaths, 276 were due to pulmonary carcinoma of the lung.[365]

Today, more than a half century later, while mining operations have become more sophisticated and dominated by the major multinational corporations, the death rate from cancer has remained largely the same. This fact can be seen through examining the largest uranium producing region in the world – northwest New Mexico. This region alone produces more than 50% of all the uranium used in the US. About 47% of the uranium in this region lies buried under American Indian land, some in areas that have been held sacred by the Indians for thousands of years.

To take one example, the Gulf Oil Corporation is currently sinking two of the world's deepest uranium shafts into the side of Mount Taylor, one of the four sacred mountains of the Navajo and Pueblo tribes, who consider it the southern boundary of their universe. To appreciate what this means to these people, it might be well to consider the reaction of Christians to sinking similar mine shafts into the sides of the Mount of Olives outside Jerusalem where Jesus taught and was betrayed. For Jews, a similar act would be to sink the shafts into the side of Mount Sinai where Moses received the Ten Commandments, or next to the Wailing Wall on the sides of Mount Moriah in Jerusalem.

Nevertheless, Gulf Oil continues to dig its two shafts, mainly because it is supported by the US government, which has actively encouraged the leasing of Indian land for mining purposes. Indeed, as of 30 June 1974, there were 380 uranium leases on Indian lands besides the one at Mount Taylor. Owning these leases, leased out by the Bureau of Indian Affairs under the auspices of the US Department of the Interior, are the following companies: Kerr McGee Corporation, the largest uranium producer in the world, Continental Oil, Mobil Oil, Humble Oil, Marathan Oil, Exxon, Anaconda, Grace, Gulf Minerals, Homestake, Hydro Nuclear, Pioneer Nuclear, Western Nuclear, and Phillips Petroleum.[366] One can see from this list the dominance of the oil corporations in the uranium mining business. According to a study by the Oil, Chemical and Atomic Workers' International Union released in April 1979, nine oil companies control 74.5% of all the uranium reserves in the US. Kerr McGee and Gulf Oil control 52% between them. Foreign interests are even coming in to exploit the rich Indian uranium resources, most particularly the Mitsubishi Oil Corporation of Japan.

The experience of the uranium miners can be seen from the examination of the mining practices of the Kerr McGee corporation which has been longest in the area of north-western New Mexico. In the 1950s, it opened up a series of uranium mines and a uranium mill near Shiprock, New Mexico, the largest population centre of the Navajo nation. Because there are no taxes on the Indian reservations and virtually no regulation of mining practices, Kerr McGee was left to exploit the land and the cheap Indian labour as it pleased. While in Europe, therefore, uranium mines were equipped with ventilation shafts to blow the radioactivity out of the mine into the atmosphere, thus lowering the cancer death rate by as much as 50%, Kerr McGee refused to put the ventilation shafts in. While federal regulations require at least an eight-hour wait after blasting before the miners are sent down the mine shafts, Kerr McGee sent its Navajo miners down the shafts immediately after blasting, even though in mines it held outside the reservation on public lands, it adhered to the eight-hour wait regulation. And finally, while miners generally were being paid union rates, Kerr McGee paid the Navajo miners only $1.60 per hour. As one of these miners recalls the sixteen years he worked in the mines: 'They chased us in (the mines) like we were slaves. I remember that it used to be so dusty that we were always spitting up black stuff and how when we went home we all had headaches from breathing all that contamination.'[367]

According to LaVerne Husen, Director of Public Health Services in Shiprock,

> Those mines had 100 times the radioactivity allowed today. They weren't really mines, just holes and tunnels dug outside into the cliffs. Inside the mines were like radiation chambers, giving off unmeasured and unregulated amounts of radon. . . . It was a get-rich quick scheme that took advantage of Navajo miners who didn't know what radioactivity was or anything about its hazards.[368]

When the mines' uranium was exhausted in 1968, Kerr McGee abandoned them, leaving radioactive build-up and over seventy acres of uranium waste. Beginning in 1970, the Navajo miners began to die. Of the 100 miners involved in the Shiprock mines, 25 were dead of 'mountain fever' by 1978, and 20 more are in the process of dying, according to Lynda Taylor of the Southwest Research and Information Centre.

This death rate becomes even more startling when it is realized that among the Navajo lung cancer is virtually unknown. In a seven-year study completed in 1972, not a single case of lung cancer was found among over 50,000 Navajos examined. Several of the widows of the dead miners as well as wives of miners still alive but dying went to Kerr McGee and requested compensation enough to keep their large families fed and clothed. Kerr McGee refused, its spokesperson Bill Phillips stating to the press: 'I couldn't possibly tell you what happened at some small mines on an Indian reservation. We have uranium mining interests all over the world.'[369]

When one considers that the nuclear industry expects to have 18,400 underground uranium miners and 4,000 aboveground uranium miners by 1990 in the US alone, the experience of the Navajo miners causes concern for the thousands of others, for even if through the installation of ventilation systems the cancer death rate is reduced by 50%, cancer among uranium miners will still be above the national average.

It is also important to be aware that the ventilation of the radon-222 out of the mines, while sparing many of the miners, does not mean that its lethality disappears. Quite the contrary. The NRC estimates that 4,060 curies of radon-222 are released from the uranium mines into the atmosphere each year for each nuclear reactor the uranium goes to. This commits the US fuel cycle to releasing 284,000 curies annually into the atmosphere of the western part of the country. If 150 reactors are allowed to operate, 609,000 curies of this gas will be released; if 500 reactors, 2,000,000 curies will be released.[370] As will be pointed out in the next section, this radon-222 stays radioactive for long periods of time.

Because of the population distribution of uranium mines in the US, the casualties will be primarily among the poor rural farmers and the American Indians on whose land most of the uranium is found.[371]

2. Milling

The purpose of the uranium mill is to extract the uranium from the ore it is mined in. This is done by crushing the uranium ore so that the rock can be dissolved by an acid or alkali leach. The uranium, mainly U_3O_8, known as *yellowcake*, is then purified and finally packaged for shipment.

The radioactivity associated with milling comes from the natural uranium and its daughter products as well as other radioactive emitters in the ore. 91.7% of the uranium is extracted from the ore; the remaining 8.3%, along with 97% of all the radioactive daughter products, are left in the ore and disposed of as *tailings*. A typical mill will produce about 800 metric tons of these radioactive tailings per day amongst some 2,500 metric tons of waste.[372]

Because the levels of radioactivity in these tailings are so 'low', according to the threshold theory, most is taken to deserted areas and simply piled up to form literally mountains of tailings. Sometimes the tailings have been used to construct earth dams. Some have been buried in abandoned mines and open pits and covered with earth.

Because the radiation emissions from the tailings were far below the allowable dose limits, a good deal of them have been used in the construction of homes, schools, buildings and hospitals as fill beneath the foundations. As there are a total of over 10 million tons of tailings in four different parts of the Navajo reservation, entire communities of Indians have used them to construct their hogans and schools. Even as far away as Grand Junction, Colorado, thousands of homes used the tailings throughout the 1960s. Because the tailings were already ground up, the building contractors saved considerable money using them.

Despite this attitude, uranium mill tailings are dangerous. Dusts containing uranium and its daughter products, thorium-230 and radon-222, are released from the ore piled outside the mill. A radioactive dust containing U_3O_8 is also released in the purifying and packaging of the yellowcake, most of which is simply blown out of ventilation shafts into the outside air. Both types of dusts, however, are insoluble aerosols. If they are inhaled, aerosols tend to remain in the lung and can cause cancer and other degenerative diseases. Radon gas is also released at all stages of the operation.

In 1976, Dr R. O. Pahl pointed out that thorium-230 in mill tailings decays to radium-226 which in turn decays to radon-222. While regular radon-222 has a half-life of only minutes, radon-222 derived from thorium has a half-life determined by that of thorium: 80,000 years.

Once the radiobiological impact of these radioactive elements is calculated in terms of their half-life instead of merely by the life-span of the mine or mill, a referent the government and the industry usually use, the figures in expected human deaths become quite staggering. According to Dr Walter H. Johnson, former Associate Director of the Oak Ridge National Laboratory in testimony before the Atomic Safety Licencing Board, on 21 September 1977, the loss of life among future citizens from each year's commitment to the nuclear fuel cycle is 28,000 from the mill tailings alone. This NRC estimate means that by the year 2000, the annual commitment to milling uranium will mean 200,000 deaths in the US.[373]

Dr R. L. Gotchy, a member of the NRC staff, estimates that active mining and milling of uranium results in 0.11 to 1.2 cancers and 0.036 to 0.40 genetic defects per reactor year for the first 100-1,000 years. This means in the US 270 to 3,000 cancers and 90 to 1,000 genetic defects at the present nuclear fuel-cycle commitment and up to 2,000 to 21,000 cancers and 630 to 7,000 genetic mutations by the year 2000, if the NRC projections for the expansion of nuclear power and the necessary mining and milling expansions are adhered to.[374] One should keep in mind that this estimate only applies for the first 1,000 years. The radon-222 has a half life of 80,000 years. The NRC has not calculated for this yet.

They are only now waking up to the fact that the tailings under the foundations of thousands of homes, schools and hospitals are emitting the same type of radon-222 that causes cancer in miners. After much debate on who would foot the bill for the removal of the tailings to try to stem the rising cancer rates that were occurring, the federal government through the Environmental Protection Agency finally agreed to do it. The project is at this moment under way, particularly in Grand Junction, where much of the tailings were used.

While the government is removing the tailings from the foundations of homes in Grand Junction, Colorado, however, nothing is being done to remove them from the homes and schools of the Indians. According to Oscar Sloan, a Navajo and former miner living near Monument Valley, Arizona, 'All the people here have used the uranium wastes to build our houses. The company never told us they were dangerous. Some white men came here a couple of years ago, and said we shouldn't live in our houses. They said the Government would get us new houses because our homes are radioactive, but they never did. I don't want to live in this house any more, but I have no place else to stay, no place else to go.'[375]

3. Uranium Conversion

In order to make the uranium yellowcake produced by the mills useable in reactors, it must be converted to uranium hexafloride, UF_6, a gaseous compound used in the uranium enrichment process. There are currently two conversion facilities in the US: a Kerr-McGee plant in Gore, Oklahoma, and an Allied Chemical plant in Metropolis, Illinois. The combined capacity of these two plants can produce between 17,300 and 21,900 metric tons of UF_6 per year.

There are two types of processes: one known as 'dry', the other known as 'wet'. In both processes, however, the radioactive releases are essentially the same, and come in solids, liquids and gases. Gaseous effluents contain $(NH_4)_2U_2O_7$, UO_2, UF_4, UF_6, U_3O_8, and UO_2F_2. Two thirds of these are released in an aerosol form.

Radioactive solids are produced by the ton: it is estimated that by the year 2,000 from 97,000 to 127,000 metric tons will have been produced containing 21,000 to 28,000 curies of radiation.[376]

Radioactive liquid emissions are estimated to range from 340 to 450 curies up to the year 2000, involving 1.7 metric tons of liquid waste from each 10,000 metric tons of yellowcake converted.

According to the US Environmental Protection Agency and the NRC, an individual living next to a wet uranium conversion plant may receive 2.9 millirems of radiation to the lung and 0.2 millirems to the bone per facility year through *inhalation*. That same individual is also allowed to receive 19.2 millirems of radiation to the bone, 3.9 millirems to the kidney, 1.6 millirems to the gastro-intestinal tract, and 3.5 millirems to the whole body through *drinking* radiated water. Someone living next to a dry conversion plant can receive 7.2 millirems of radiation to the lung and 0.5 millirems to the bone through inhalation and 50.8 millirems to the bone, 1.4 millirems to the kidney, 1.6 millirems to the gastro-intestinal tract, and 33.1 millirems to the whole body through drinking the radiated water downstream from the plant.[377] All these figures have been calculated in accordance with the threshold theory.

Unfortunately, no studies have been made of the communities around the conversion plants to determine the effects of all this 'allowable' radiation.

4. Uranium Enrichment

Raw uranium hexafloride after the conversion process contains less than 1% uranium-235 and more than 99% uranium-238. Reactors must use uranium fuel that is from 2% to 4% uranium-235, however, and so enrichment plants 'enrich' the uranium to the desired percentage level of the -235 isotopes. There are three such plants in the US and one in France. Britain, West Germany and the Netherlands are now in the process of constructing a gas centrifuge enrichment plant at Capenhurst, England, and at Almelo, Netherlands. These enrichment plants are enormous both in size and in the amount of energy required to carry out the enrichment process. For example, the single plant at Oak Ridge, Tennessee, covers over 150 square acres of building area and requires more electricity to operate than the cities of Nashville, Knoxville and Chattanooga with their surrounding populations combined.[378]

In the enrichment process, 0.007% of all the uranium is released directly into the environment either as a gas or liquid effluent. This is allowed, however, because at the enrichment stage of the fuel cycle the NRC and the US Environmental Protection Agency have again set 'allowable' dose limits for radiation exposure: 0.05 curies per person per year per facility if inhaled and 0.6 curies per person per year per facility if taken in through drinking water.[379]

Occupational dosages are also high. The NRC estimate of the annual dose of radiation exposure through inhalation alone among the workers at the plant is about 700 millirems per year. This estimate is based on routine emissions allowances.

The solid waste amounts to 100 metric tons of uranium sludge tailings per year, most of which is buried on site, although the uranium-238 solids are stored for future use. The radioactive emissions and effects of the tailings in the enrichment stage are similar to those at the milling stage.[380]

Statistics relating to increases in cancer and leukemia rates can be given for enrichment facilities; they are similar to increases for the other aspects of the nuclear fuel cycle. Rather than give statistics, however, I shall highlight a well documented case of one enrichment worker who in many ways exemplifies the point being made. This is the case of Joe Harding who worked at the

US government's uranium enrichment plant at Paducah, Kentucky, during the 1950s and 1960s.[381]

Harding began work with top-secret security clearance in 1952 and was assigned to work as a process operator in a building the size of five football stadiums. There, he and other workers kept constant watch on some of the 4,000 process stages needed in the enrichment process, plugging leaks and pulling equipment off the line for repairs. He also worked in the Product Withdrawal Room, where the pressurized uranium gas is drawn off, analysed for its isotopic assay, and pumped into 14 ton cylinders for shipment to Oak Ridge, Tennessee, if further enrichment is needed.

Releases of radioactive gases were routine, Harding recalls, as was the constant release of uranium dust: 'At the end of a day you could look back behind you and see your tracks in the uranium dust that had settled that day. You could look up at the lights and see a blue haze between you and the light. And we ate our lunch in all this, every day, eight hours a day. We'd just find some place to sit down, brush away the dust, and eat lunch.' Harding did not worry about the dust, however, recalling that he, 'just like everyone else believed the company when they said we would be exposed to no more radiation than we'd get from wearing a luminous-dial wristwatch. They said they'd protect us and let us know if we ever got too much and we all believed them.'

One of the ways the company (Union Carbide ran the plant for the government) protected the workers was by giving them film badges and taking urine samples. But Harding remembers several incidents where something abnormal happened but nothing was ever done in response: 'We had these film badges we wore to indicate exposure. Every few days they'd take up the badges and send them off to the Oak Ridge National Laboratory (also run by the government by Union Carbide) for analysis. One day a few of us men laid our badges on a smoking chunk of uranium for eight hours and turned it in. We never heard from it. They took urine samples from us every ten days. Once somebody dropped a small chunk of uranium in the urine sample. Nothing was ever said about it.'

Harding further remembers that falsifying records concerning radiation exposure was a common practice encouraged by his supervisors:

When we took radiation surveys of the Product Withdrawal Room and the cylinders (of enriched uranium gas) we zeroed the meters right there in the room, right where all the radiation was to begin with. It never dawned on us that we were zeroing all this out and only showing additional radiation, which we did find.

Some of the cylinders would accumulate 'heels' of uranium metal coating on the inside, adding as much as fifty pounds to their weight. Sometimes these heels could make the cylinders pretty hot, and when we'd get near them with our meters, the meters would go off scale.

We began to realize that the radiation surveys were useless because we were zeroing the meters right in the area and we were told to ignore the heels.

We had to log all the readings – ten feet above a cylinder, ten feet to the side – and my first experience in this was with a cylinder that had a high heel in it. It was too hot to meet acceptable levels for shipment. I was on the midnight shift and I called over to the Roundhouse (the control room) and I said to the shift supervisor, 'Hey buddy, I've got a cylinder over here in No. 2 position, and we can't ship it, it's too hot.'

The supervisor said, 'What do you mean you can't ship it? See if you can't get me a better reading.' So I was just learning, but I thought, 'Hey, I've read this thing once.' So I read it again with two or three different meters. Same story. I called back and told them, 'I just can't ship it, it's just too hot.' So the shift supervisor said, 'Listen, we've got these cylinders in the sequence that they were withdrawn and that's the way we want to ship them. We've already got this cylinder reported to go on this truck. So let's get a good reading and get it on the truck.'

So he's the boss and it then dawns on you that he means just what he said. So you just say to hell with what it reads and you put down an acceptable reading and you ship it.

After an occasional two or three instances like that and some talking around you find out everybody is doing the same thing. The radiation meters sat over in the corner and got cobwebs on them.

We all thought, why fight it? You're going to lose your job if you don't go along with it. It was the same way with the purge rate (the amounts of radiation that were routinely and legally released from the plant). And if we had tanks of con-

> taminated liquid or gases that had to be disposed of, we'd just wait till a dark night when there was no moon and just shoot it right up the stack. Sometimes we didn't even wait for a dark night.
>
> Of course, we were dumping it around on all these farms, but that was okay as long as nobody knew, and after a while we stopped using this elaborate pumpdown procedure altogether.
>
> But nobody ever told you what to do, they just didn't question it. They'd tell you, 'Clean 'em up.' And if you cleaned 'em up, 'That's a good job, bud, you're a fine man.' Well, they knew what was being done with it – that was the way they designed it.

Within a few years of working at the plant, Harding began to suffer violent vomiting attacks and weight loss. Finally, in 1961, his stomach was removed. The organ was displayed in formaldehyde at the hospital, so advanced was its state of deterioration. Since 1953, small skin sores have been working their way up Harding's body, and beginning in 1968 he has suffered bouts with pneumonia every single year. Moreover, since 1971, fingernail-like growths of cartilage have been penetrating the skin over Harding's toes, ankles, wrists, finger joints and knuckles. Similar growths have developed from the ends of his lower ribs as well, forming hard knots beneath his skin. Each morning upon rising he spends two hours digging the spur-like growths out of his feet and hands with a knifeblade. Since 1978, Harding has also developed a slight palsy.

According to Dr William Jennings, a neurologist at the Western State Mental Institution in Bolivar, Tennessee, who examined Harding in 1979, 'the obvious anatomical findings probably were caused by radiation exposure. Nothing else was found to indicate any other cause.' His conclusion: 'Joe Harding's whole medical history certainly smacks of radiation damage. The things that are wrong with him could not be proven on any other organic basis.'

Harding has not allowed himself to be overcome by the debilitation of his body, however, and like those few Hiroshima survivors who were capable of positive formulation and survivor mission, he has taken it upon himself to find out what has happened to each of the 200 men who began working with him back in 1952. Part of his list includes the following: 'Wade

McNabb, died in 1972 at age fifty-five. He first came down with it in 1958. Leukemia. And Eugene Ragland, forty-seven years old, died in 1978 of stomach cancer and leukemia. Dave Wilson, stomach cancer, 1978. Jack Owens, died in 1956 with cancer. He was in his thirties, so young. Carroll Groves, 1974. One night he laid down and went into a coma. Died a few days later.'

Says Harding: 'The evidence just keeps mounting. Every day I hear about one or two more that's died.'

Joe Harding himself died on Saturday 1 March 1980. Although the official death certificate asserts that his death was due to heart failure, the autopsy revealed a thirty pound cancerous tumour spread around his spinal column and throughout his back muscle tissue.

5. *Fuel Fabrication*

When the uranium has been enriched to its desired percentage level of uranium 235, it is shipped to a fuel fabrication facility where the UF_6 is transformed into a dioxide powder, which is then pressed into small pellets, sintered, and finally loaded into stainless steel or zircalloy tubing, called cladding, twelve feet long by half an inch wide (3.7 meters by 13 mm). These are the fuel rods, 36,000 to 50,000 of which go to make up the reactor core of each nuclear power plant.

At this stage radioactivity is again released into the atmosphere and water, and 'allowable' dosages are set for both workers and the public.

One particular fuel fabrication plant that has received attention recently is the Getty Nuclear Fuel Services plant just up the Nolichucky river from Jonesboro, Tennessee. The plant began operations in 1958 and was contracted to produce both mixed-oxide and enriched uranium fuels for the Navy's nuclear programme.

By 1973, after the fifteen-year latency period for many cancers, Jonesboro began to experience a sharp increase in its cancer rate. A State Health statistician, Tom Brian, graphed these increases and found them to be dramatically higher throughout Unicoi County, where Jonesboro is located, than for the rest of the state. In comparing this increase with national trends, it was discovered that Unicoi County had the third most rapidly increasing cancer rate in the US. By 1978, the *Atlanta Journal* reported that in the twenty years the plant had been operating the county death rate from cancer had increased 100%.[382]

This could be due in part to such occurrences as the 1977 'routine' emission into the Nolichucky river of between 250 and 500 pounds of enriched uranium. Plant releases had continually exceeded state limits for fluoride, mercury, solid organics and nitrogen as well, but this negligence was not prosecuted by the State of Tennessee or punished by the NRC.

In 1978, river sediment samples were collected five miles downstream from the plant. 0.14 picocuries per gallon of plutonium-238 were discovered. Tests were also taken of the Jonesboro drinking water. Positive readings of radioactive heavy metals were found ranging as high as 3.2 plus or minus 1.6 picocuries per litre.

At this point, the townspeople complained. In response, the NRC sent three officials from its regional office in Atlanta, Georgia, to address a townsmeeting in May, 1978. The NRC officials informed the people that it took a 50,000 millirem dose to cause cancer and that the fuel fabrication plant was only giving 0.2 millirems to each of the townspeople. They further assured the audience that not one single facility licensed by the NRC had caused either death or illness to anyone living nearby. The people expressed disbelief at these words but when the meeting was over the NRC officials went back to Atlanta. No further contact with Jonesboro was made; no action was taken against Getty Nuclear.

Besides the risk of radiation exposure from enrichment plants, there is another element of the process that gives rise for concern. This involves the stainless steel and zircalloy tubing in which the uranium pellets are placed. They are composed in part of rare metals such as chromium and zirconium, which are virtually now completely depleted as natural resources. Despite this fact, there is no attempt at conservation. For every metric ton of fuel consumed at a nuclear reactor, 1.4 metric tons of stainless steel and zircalloy are required. This cannot be recycled, for in the fissioning of the uranium pellets inside the tubing, the tubing itself becomes so radioactive that when the fuel is spent, the tubing must be buried as radioactive waste. We are thus throwing away thousands of tons of these rare metals each year, simply to be able to generate electricity in a way that will not be ultimately needed.

6. Nuclear Reactors

From the fuel fabrication facilities the fuel rods are sent to the nuclear reactors. Here they are placed in the reactor core and the process of fissioning the uranium-235 in the pellets takes place. This fissioning produces heat which turns a turbine connected to a generator – which produces electricity.

Among the more important radionuclides produced when the uranium-235 is fissioned are isotopes of kryption and xenon gases, cesium and rubidium metals, alkaline earths such as barium and strontium, and halogens such as iodine and bromine.

After the neutrons are fissioned from the uranium-235, most are absorbed into the uranium-238 which composes over 95% of the fuel rod pellets. It is at this stage that the most dangerous radioactive agent of all is produced, for when the uranium-238 absorbs an extra neutron it becomes uranium-239, which quickly degenerates into a daughter product: plutonium-239. Other radioactive products are formed as well: argon-41, fluorine-18, nitrogen-13, nitrogen-16 and oxygen-19.

Reactors are constructed with multiple barriers in order to isolate and contain as much of this radioactive release as is possible. The principle barriers are the fuel cladding around the uranium pellets, the covering around the reactor itself and the containment building in which the reactor is placed.

Penetration of these three barriers can and does occur. For example, the fuel rods are only generally 25/1000ths to 33/1000ths of an inch thick. With all the stresses of heat and reactor operation, splits and holes are made, allowing radioactivity to escape into the coolant. In fact, so inevitable are the fracturings of the fuel rods that the reactor system containing them is designed to accommodate a 1% defect rate, meaning that 1% of their radioactivity is allowed to be released into the coolant.

The other systems are equally vulnerable and are similarly designed for minimal rates of defects and consequent radioactive releases. Again, it should be recalled that this is allowed because the threshold theory assumes that certain amounts of radioactivity are tolerable. The following tables give *planned* airborne releases for *normal* reactor operations:[383]

Representative Airborne Radionuclides and Planned Release Rates

Environmental Protection Agency, 1973

		^{3}H, Ci	^{14}C, Ci	^{85}Kr, Ci	Radionuclide	Average stack release
Accumulated radionuclides discharged immediately after reactor shutdown	=	0.17	0.11	<0.0005	^{54}Mn	5 pCi/sec
					^{60}Co	8
					^{90}Sr	8
Radionuclides discharged continuously during refuelling	=	13.	0.066	2.8	^{131}I	<9

Planned Airborne Release Rates for Normal Reactor Operations

Estimated Air Ejector Off-gas Release Rates Following 30-Minute Holdup 1064-MWe BWR with 0.25% Failed Fuel, 18.5 scfm in-leakage

Radionuclide	Emission rate μCi/sec	Annual Discharge Ci/yr
Nitrogen-13	340	8,580
Krypton-83m	2,537	64,000
Krypton-85m	5,700	143,800
Krypton-85	7.5	189
Krypton-87	15,700	196,000
Krypton-83	17,367	438,000
Krypton-89	262	6,610
Xenc ,-131m	15	378
Xenon-133m	188	4,743
Xenon-133	5,100	128,700
Xenon-135m	8,000	202,000
Xenon-135	17,367	438,150
Xenon-137	860	21,700
Xenon-138	26,500	668,600
Total	100,000	2,523,000

1% Failed Fuel gives 4 times these values (10,092,000 Ci/yr).

Source: Browns Ferry Final Safety Analysis Report.

Worldwide Health Risk Contributions from Reactor Tritium Releases

TRITIUM Treatment Option	Annual Discharge (Ci/yr)	Total Year 2000 30-Yr Commitment For All US Plants Person-Rem	Health Effects
BWR-1,2,3	200	1,080	0.78
BWR-4	130	702	0.50
PWR-1,2	1200	12,960	9.3
PWR-3,4	760	8,208	5.9

Estimated Reactor Coolant Specific Fission Product and Corrosion Product Activities (at 578°F)
Fission product reactor coolant concentrations corresponding to 1% Failed Fuel

Isotope	μCi/cc	Isotope	μCi/cc
Noble Gas Fission Products		**Fission Products**	
Kr-85	1.11	Br-84	3.0×10^{-2}
Kr-85m	1.46	Rb-88	2.56
Kr-87	0.87	Rb-89	6.7×10^{-2}
Kr-88	2.58	Sr-89	2.52×10^{-3}
Xe-133	1.74×10^{2}	Sr-90	4.42×10^{-5}
Xe-133m	1.97	Y-90	5.37×10^{-5}
Xe-135m	0.14	Y-91	4.77×10^{-4}
Xe-138	0.36	Sr-92	5.63×10^{-4}
Total Noble Cases	187.3	Y-92	5.54×10^{-4}
		Zr-95	5.04×10^{-4}
Corrosion Products		Nb-95	4.70×10^{-4}
Mn-54	4.2×10^{3}	Mo-99	2.11
Me-56	2.2×10^{-2}	I-131	1.55
Co-58	8.1×10^{-3}	Te-132	0.17
Fe-59	1.8×10^{-3}	I-132	0.62
Co-60	1.4×10^{-3}	I-133	2.55
Total Corrosion		Te-134	2.2×10^{-2}
Products	3.7×10^{-2}	I-134	0.39
		Cs-134	7.0×10^{-2}
		I-135	1.4
		Cs-136	0.33
		Cs-137	0.43
		Cs-138	0.48
		Ce-144	2.3×10^{-4}
		Pr-144	2.3×10^{-4}
		Total Fission Products	12.8

Source: Kewaunee Final Safety Analysis Report, Appendix D, Table D 4-1

Primary-to-Secondary System Leakage Experience in Pressurized Water Reactors

Plant	Average Leakage Rate (gallons) per day)	Duration
H. B. Robinson	55	7 months
Unit 2	14,400	<1 day
Point Beach Unit 1	up to 50	several months
Connecticut Yankee (Haddam Neck)	1,500	several days
San Onofre	up to 15	several months
Unit 1	up to 95	several weeks
Yankee Rowe	up to 1,200	several months

Source: Operating reports for these respective plants.

Representative Estimated Gaseous Releases
Associated with Primary-to-Secondary
Leakage (20 Gallons per Day)

Principal Radionuclide	Annual Activity Release to the Environment (curies per year) from		
	Containment Purge	Waste Gas Processing System	Steam Generator Leakage
Krypton-85	13.0	791	2.0
Krypton-87	0.04	–	3.0
Krypton-88	–	–	10.0
Xenon-131m	10.0	63	3.0
Xenon-133	1005.0	1500	682.0
Xenon-135	0.018	–	3.0
Xenon-138	0.007	–	2.0
Iodine-131	0.018	–	0.62

Source: (Table III-3) of the AEC Draft Environmental Impact Statement for Indian Pt 2

The table on page 187 indicates which parts of the body are susceptible to these planned radioactive releases and what the principal exposure pathways are.[384]

The first study attempting to correlate the radiation released from nuclear reactors with adverse health effects was conducted by Dr Ernest Sternglass in 1973 of the Shippingport reactor in Pennsylvania. Measurements of the radioactivity of the rivers which were used to cool the plant were carried out under the Surface Water Quality monitoring programme by the Pennsylvania State Department of Health. In 1964, and again in 1970, the total amount of beta radiation measured one mile below the plant exceeded by many times the beta activity measured 25 miles above the plant. The amount of radioactivity dumped into the Ohio river altogether in 1964-5 was measured at 78 curies; in 1970, 182 curies. The State Health Permit limit was only 0.58 curies per year; the AEC limit for 'maximum liquid discharges' was 22.6 curies per year.

The company operating the reactor produced documents indicating that its discharges had been below acceptable limits: 0.53 curies in 1964, 0.14 curies in 1965, and 0.07 curies in 1970 – some 2,500 times less than the findings of the State Department of Health.

In 1971, the NUS Corporation, in an environmental survey of the sediment of the Ohio river, found 300-400% excess radioactivity concentrations of strontium-90 and cesium-137 in the milk of cows grazing in pastures within ten miles of the plant as compared with milk sampled elsewhere.

Representative Exposure Pathways for Radiation Exposure from Reactor Effluents

Environmental Protection Agency, 1973

Radionuclide	Discharge Mode	Principal Exposure Pathways	Critical Organ
Radioiodine	Airborne	Ground deposition – external dose	Whole body
		Air inhalation	Thyroid & Embryo
		Grass cow-milk	Thyroid & Embryo
		Leafy vegetables	Thyroid & Embryo
	Water	Drinking water	Thyroid & Embryo
		Fish	Thyroid & Embryo
Strontium	Airborne	Ground deposition – external dose	Whole body
		Air inhalation	Bone and whole body
		Grass cow-milk	Bone and whole body
		Leafy vegetables	Bone and whole body
	Water	Drinking water	Bone and whole body
		Fish	Bone and whole body
Tritium	Airborne	Air inhalation and transpiration	Whole body
		Submersion	Skin
	Water	Drinking water	Whole body
		Food consumption	Whole body
Noble Gases	Airborne	External irradiation	Skin
		Air inhalation	Whole body
Cesium	Airborne	Ground deposition – external dose	Whole body
		Grass cow-milk	Whole body
		Grass-meat	Whole body
		Inhalation	Whole body
	Water	Sediments – external irradiation	Whole body
		Drinking water	Whole body
		Fish	Whole body
Transition Metals (Fe,Co,Ni,Zn,Mn)	Water	Drinking water	G.I. Tract
		Fish	G.I. Tract
Direct vector radiation		External irradiation	Whole body and skin

After gathering together all the data on radioactivity levels Dr Sternglass then gathered all available data on cancer and leukemia rates. He concluded that as a result of radioactivity released from the Shippingport reactor, leukemia and cancers of the lymphatic and blood-forming systems of the body rose nearly 70% between 1957, when the plant opened, and 1967. In 1968, cancers from all causes were 30% higher than the cancer death rate had been before the reactor began operations. During the same time, cancer rates rose in the state as a whole by only 9% Dr Sternglass also observed that the Pittsburgh area milk, which came largely from the irradiated pastures around the plant, contained 50% more strontium-90 than found in milk sampled elsewhere. Children dying from leukemia in this area had risen 50% by 1972.[385]

The rise in cancer rates alone can be seen from the following chart:[386]

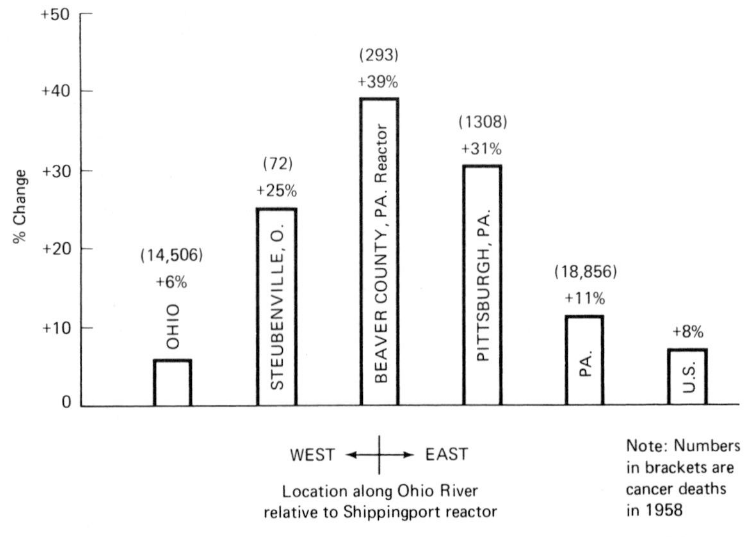

Percentage Change in Cancer Mortality Rate (1958–1968) with distance away from Shippingport Nuclear Reactor

Dr Sternglass then conducted another correlation study of the population around the Millstone I reactor in Connecticut, which in 1975 had discharged 2,970,000 curies of radioactivity into the river used to cool it. The results were similar: increases in radioactivity around the plant and corollary increases in cancer and leukemia rates. The chart on page 189 indicates the rate of cancer increases in communities around the plant both before the plant began operations (1970) and after five years (1975).[387]

The utility companies owning the Shippingport and Millstone reactors, as well as the atomic industry and the federal government, all attacked Dr Sternglass both professionally and personally for even conducting such studies. Whatever radiation

Cancer Death Rate per 100,000 Population

	Approx. Dist. from Millstone	1970	1975	Percent Change
Vermont	200m. NW	176.1	173.9	− 1%
Connecticut	35m. NW*	168.1	188.4	+12%
New Haven, Conn.	30m. W	200.9	255.5	+27%
Waterford, Conn.	0	152.6	241.8	+58%
New London, Conn.	5m. E	177.4	255.0	+44%
Rhode Island	50m. NE	200.1	216.0	+ 8%
Massachusetts	70m. NE	185.0	198.4	+ 7%
New Hampshire	120m. NE	180.4	182.6	+ 1%
Maine	200m. NE	197.7	185.0	− 6%
US	–	162.0	171.7	+6%

* Population centre of State of Connecticut (Hartford-Waterbury area)

releases there had been, they said, were all within allowable limits, and any strontium-90 in the milk was due to the fallout from the atmospheric nuclear weapons tests in the 1950s. According to Allan Richardson, Chief of the US Federal Guidance Branch in the Office of Radiation Programmes of the Environmental Protection Agency,

> I think as we all know, the doses to the average member of the population from nuclear power are all on the order of 1 millirem or less with a few minor exceptions that don't go much above that.[388]

These types of statements are indicative of the government and nuclear industry responses to credible scientific studies demonstrating adverse health and environmental effects from operating nuclear power plants. Nevertheless there have been other studies besides those of Dr Sternglass that substantiate the claim that operating reactors releasing 'allowable' amounts of radiation are causing severe damage to both the people and the ecology. Abnormal spills of radioactivity obviously cause even more damage as the case of Millstone I indicates.

Studies in Wisconsin, South Carolina and other states have all shown that there is a corollary rise in the strontium-90 level in the milk of dairies when there are recorded releases of radiation from nearby reactors. The NRC is at least consistent here as in the case of the Sternglass studies and blames the rise in strontium on the Chinese nuclear tests. This is like saying there is a huge umbrella over the American sky with holes in it just over nuclear facilities.

The Wisconsin study is worth considering in more detail.[389] It is based on fourteen years of milk monitoring by state officials in the heart of America's dairyland. A chain of fourteen nuclear reactors border the area on three sides. The study began in 1963 when a Wisconsin environmental group asked the State Radiation Section to prepare dose estimates based on the official milk sampling programme. The state refused. So the group of environmental professionals – the Land Educational Association Foundation Inc. – took on the project themselves with the help of a biology professor from the University of Minnesota. They used only data from the state monitoring records.

The chain of reactors grew from two to fourteen between 1970 and 1976. In 1973 state records show that the strontium-90 levels jumped from just below the national average to more than twice the national average. This new high level continued until 1976 when the study had to be discontinued because the State refused to show what were supposed to be public records.

The largest increases in the strontium came in an area down wind and down river from Minnesota's Monticello reactor and in the Green Bay area below the Point Beach reactor. Both high readings came and persisted during the year that Monticello was leading the nation in gaseous releases of radioactivity and Point Beach was tripling its allowed emissions.

Since the Wisconsin study only looked at three of the several dozen radioactive products produced by the reactors – strontium-90, cesium-137 and iodine-131 – they were careful to state that their findings were not the entire story. Using government formulas they assert the average yearly dose to Wisconsin citizens from radiation in the food and the environment to be 33 millirems of whole body radiation for an adult and 67 millirems for a growing child. This is significant since on 1 December 1979, the Environmental Protection Agency set the dose limit at 25 millirems. The Wisconsin study has found that the yearly dose to the bones of adults is 76 millirems and 174 millirems to growing children. This dosage of radiation has more than doubled the risk of blood cancer for Wisconsin fourteen year-olds.

The study goes on to caution that radiation comes from much more than just milk. While milk is easy to monitor it is in fact one of the least radioactive of all the foods we eat, containing only 10% of the radioactivity found in the grass eaten by the cows. Foods high in fallout radiation include potatoes, whole wheat, leafy vegetables, soyabeans, cabbage, berries, nuts, and cheese. Cheese multiplies the radiation dose of milk six times.

The NRC rejects the Wisconsin study despite the fact that it is based on verified state records and calculated with NRC and EPA formulas. One NRC official stated that the study 'showed extreme bias in its data and its presentation when we reviewed it'. Indeed, the NRC is stopping the monitoring of strontium-90 around US nuclear reactors. The reason given is that the utilities, asked by the NRC to monitor themselves, are not finding much.

In Europe meanwhile a new report written by a team of West German scientists – physicists, agricultural biologists, chemists, and mathematicians – from the University of Heidelberg confirms the conclusions of Dr Sternglass and the Wisconsin study. Called the Heidelberg Report, it asserts that normal releases from reactors are exposing people to dangerous amounts of radiation.[390] The report breaks new ground because for the first time the foundations of the NRC's safety assurances are examined. This is important for many European countries because many of their reactors are patterned after American designs, all of which were built with NRC calculations.

The NRC claims that operating reactors give off between a fraction of a millirem to less than five millirems to people living within ten miles of the plants. This amount is virtually insignificant in the light of the fact that a chest X-ray exposes a person to between fifteen and thirty millirems. The Heidelberg report asserts that this assertion is 'unrealistically low' due to the fact that the NRC judgment on how much plutonium, cesium and strontium crops pick up is between 10 and 1,000 times too low. The Report makes this claim after examining the studies upon which the NRC bases its conclusions. These are old AEC studies done back in the 1950s when the AEC was attempting to demonstrate that not only was it on top of the nuclear weapons fallout situation but that such fallout posed no danger to the public.

The AEC studies are interesting. They were conducted to determine whether food crops would take up dangerous levels of radioactivity from the soil. Before beginning the experiments, AEC scientists made preparatory tests on a variety of soils, choosing for their experiments those soils which absorbed the *least* amount of radiation. Furthermore, as it was then well known that plants have difficulty assimilating radiation until they are acted upon by soil bacteria, the scientists cooked their soil in ovens, thus *killing* its bacteria. Finally, the AEC experimenters added the radioactive substances to the soil just shortly before the plants were harvested, thereby avoiding the normal condition of plants growing from seed in contaminated soil. Not surprisingly, their conclusions indicated that hardly any of the nuclear weapons fallout was getting into crop plants. It is upon foundational studies such as these that the NRC bases its calculations for 'allowable dosages' and its continued assurances to the public that operating nuclear reactors pose little or no danger to people living close by.

After examining the NRC safety estimates, the Heidelberg Report breaks other new ground by calculating the dose from a nuclear reactor with figures chosen by independent scientists. Using these more accurate figures, the German scientists fed them into the NRC computer model. What they discovered is alarming. They found that a pressurized water reactor planned for the city of Whyl on the Rhine can be expected to expose the surrounding population to an annual dose of 1,071 millirems of whole body radiation. The major part of this dose would come from radioactivity taken into the body through food and drink. Atmospheric releases by the reactor account for most of this contamination but releases into the Rhine are important as well.

It is important to bear in mind that this 1,071 millirem exposure amount cannot be woodenly transfered to other populations around other reactors. Conditions vary enormously due to plant size, wind conditions, proximity of farm land, size of population and its distance to the reactor, and how much radiation is in fact released.

Another study done in West Germany, this one by the Bremen Institute for Biological Safety, confirms the general conclusion of the Heidelberg Report.[391] The study discovered that during the ten years before a reactor near Bremen called the Lingen reactor, began operating, the surrounding population experienced 30 leukemia deaths. In the ten years *after* the plant began operations, however, between 1968 and 1978, this same population experienced 230 leukemia deaths, 170 of which have been to children under the age of fifteen.

Besides correlation studies of populations around reactors there have been studies of atomic facility workers. The most extensive study has been that of Dr Thomas Mancuso of the University of Pittsburgh, who was asked by the AEC to conduct an investigation of the long-term health of the atomic workers at the Hanford Atomic Works nuclear complex operated by the US government and Battelle Northwest Laboratories. The complex includes several nuclear reactors and is located at Richland, Washington. Collaborating with Dr Mancuso by independently analysing his data have been two eminent British scientists, Dr Alice Stewart of the Department of Social Medicine of the Queen Elizabeth Medical Centre, Birmingham, and Dr George Kneale of Oxford University. The study involved a population of some 20,000 men and 5,000 women over a period of 29 years.

The results were astonishing. While scientists embracing the threshold theory believed that 33 rems of exposure was the threshold for observable effects on the body, Mancuso, Stewart and Kneale discovered that this amount is in fact the *doubling dose* for all cancers. Indeed, while the legal limit of five rems can be given to any worker per year, Mancuso found that the estimated doubling dose for bone marrow cancer is 800 millirems, less than even one rem; for pancreatic cancer 7.4 rems; for lung cancer 6.1 rems; for all RES neoplasms 2.5 rems.

The average dose range for the workers was only one rad per year in actual practice. Given this one rad per year dose, however, the study found that the risks for all cancers increased 26%; for RES neoplasms 58%; for bone marrow cancers 107%. Women had a higher death rate from cancer than men; about 31% of the 412 women who had died by 1973 died of cancer, well above the national average of 18%. This may be due to the fact that for women the doubling dose for cancer is only 9 rads, not 33. Older men were also found to be more susceptible than young men. Those between the ages of 30 and 40 had cancer rates lower than expected, while those between 45 and 50 had a rate 15% higher than expected. Those over 50 had rates 50% higher; those over 55 had a cancer rate three times the norm; and those over 70 had a cancer rate *fourteen times* the norm.

The study also found that cancer has a relatively stable *latency* period: for all cancers 12 years; for RES neoplasms 11 years; for bone marrow cancers 9 years; for lung cancer 14 years; for brain tumours 17-19 years; for cancer of the large intestine 18 years.

7. Reprocessing

After a certain amount of time, the uranium-235 inside the fuel rods has all been fissioned and the uranium-238 has become totally saturated with extra neutrons. At this point, the fuel rods must be removed and either buried as waste or reprocessed to extract for reuse some of the radioactive elements still usable. The elements still usable are the uranium and plutonium.

Reprocessing was originally used to produce plutonium for use in nuclear weapons. Britain was capable of mounting a large-scale reprocessing of uranium fuel to extract plutonium shortly after World War II at its Windscale facilities. It was the 1957 fire that forced these two plants, one for chemical separation and the other for plutonium extraction, to be permanently entombed.

In the late 1950s the British Atomic Energy Authority decided to build a new reprocessing plant, this time designing it to accommodate not only its plutonium needs for nuclear weapons but is growing civil nuclear programme as well. This new plant came into operation in 1964. A similar reprocessing plant can be found in La Hague, France. Two operate in the US. In 1976, British Nuclear Fuels announced plans to build a second even larger reprocessing plant at Windscale. Whether this one will be built or not remains to be seen, given the widespread public opposition to it.

At the Windscale and La Hague reprocessing facilities, reactor fuels are not dealt with by the pound but by the ton – by the *hundreds of tons*. Day and night huge transporters with lead tanks travel from all over Europe – from Germany, Italy, Holland and Spain – to deposit their radioactive waste to be reprocessed. At the present time the largest contracts are coming from Japan, which, because Windscale is already working to full capacity, La Hague takes in.

Technical failures at these plants, while they should be rare, according to the planners, are commonplace. In practice, hardly an hour passes without a repair being needed somewhere in the facility. In zone 817 of the La Hague plant, for example, where plutonium oxide is treated and weighed, there was only one accident-free day during the four weeks from 23 January to 21 February 1977. The alarm was sounded over forty times, and the rooms had to be completely evacuated five times due to the fact that floor contamination reached 10,000 disintegrations per second.[392]

A particularly bad problem is ventilation. In order to keep the air relatively free from radiation or at least under the allowable limits, a very complicated system of air filtering is necessary, sometimes requiring up to 100 air changes an hour in areas such as the plutonium workshops. But the system continually breaks down, both at Windscale and La Hague, and is sometimes not corrected for hours, even days, until it is observed that the body counts of plutonium contamination are rising.

A situation similar to this apparently occurred at the Atomic Weapons Research Establishment in Aldermaston, England, during August 1978, when a considerable number of employees became infected with high doses of plutonium.[393]

The most immediate impact of all the leaks, breakdowns and ventilation foul-ups are upon the workers. From 1973 to 1975, the number of worker contaminations officially reported by the La Hague facility rose from 280 to 572. Since then, workers report that the number of contaminations have increased, but the management no longer publishes the figures.

Besides radioactive releases contaminating the Windscale and La Hague plants *internally*, however, considerable quantities of radioactive materials in both gaseous and liquid form are released into the *external* environment. At La Hague, in 1974, in addition to small quantities of mercury-203 and iodine-131, large amounts of tritium and krypton-85 were released. In 1975, 11,000 curies of tritium, 23,000 curies of ruthenium-106, 1,000 curies of cesium-134 and -137, and large quantities of strontium-90 and plutonium-239 were released into the outside environment through the water pipes.[394] At the Windscale facility the situation is similar. The last major radioactive leak alone released over 30,000 curies of radioactive liquid into the surrounding ground. This occurred during April 1979.

As a result of continuous emissions of radioactivity, the Irish Sea around Windscale and the Atlantic ocean around La Hague show above-normal levels of radiation pollution for a radius of over 100 kilometres. Tests by the French Atomic Energy Commission along the Atlantic coast around La Hague have shown measurements of radioactivity in the marine fauna more than five times higher than those for Cap Frehl, 100 kilometres away.[395] High concentrations of radioactivity are also found on sediments, corals, mussels, algae, oysters, crabs and fish. The French fisherpeople talk of wart-like spots on the skin of their perch, butts and lamprey catches. Irish fisherpeople have similar complaints.

Unfortunately, no long-term correlation studies have yet been made around either the Windscale facility or La Hague. In late 1979, however, a study done by scientists at the University of Manchester, England, revealed that there had been a doubling in the leukemia rate for all of the major towns of Lancashire, a county which is downwind from the Windscale plant. The scientists conducting the study concluded it by stating that they saw no other potential cause for the doubling of the leukemia rates than the presence of the radiation coming from Windscale. Paul Schofield, the Chief Medical Officer of Windscale, came out with a scathing attack on the study, stating that the presence of Windscale could not possibly be the reason, as its radiation releases had all be within allowable limits. His attack gives evidence that the nuclear industry is still not prepared to admit that small amounts of radiation are seriously damaging to human health.

8. Transportation

Transportation of nuclear materials whether for fuel or for the weapons programme is an essential aspect of the nuclear fuel cycle. The uranium must be shipped from the mines to the mills, to conversion facilities, to enrichment plants, to fuel fabrication plants, to nuclear reactors. After being used in the reactors the spent fuel must then be transported to either reprocessing plants or waste disposal facilities. The extent of the transportation of radioactive substances can be seen from the chart opposite.[396]

Most shipments move on regular commerce routes and in regular conventional transport vehicles. They are thus subject to the same transportation hazards as non-radioactive cargo – including accident and theft. There is virtually no special routing, escort or speed limits to the transportation of radioactive fuel. The only protection to the public is dependent upon the design of the *containers* the nuclear materials are shipped in. In some cases careful design has been insisted upon by the atomic regulatory agencies; in other cases not.

What can and does occur in the shipment of radioactive materials has been shown by an American Friends Service Committee study entitled *Transportation and Nuclear Cargo in the South*:

> On September 21 1977, a huge cask of highly enriched weapons grade material from South Africa was loaded on a

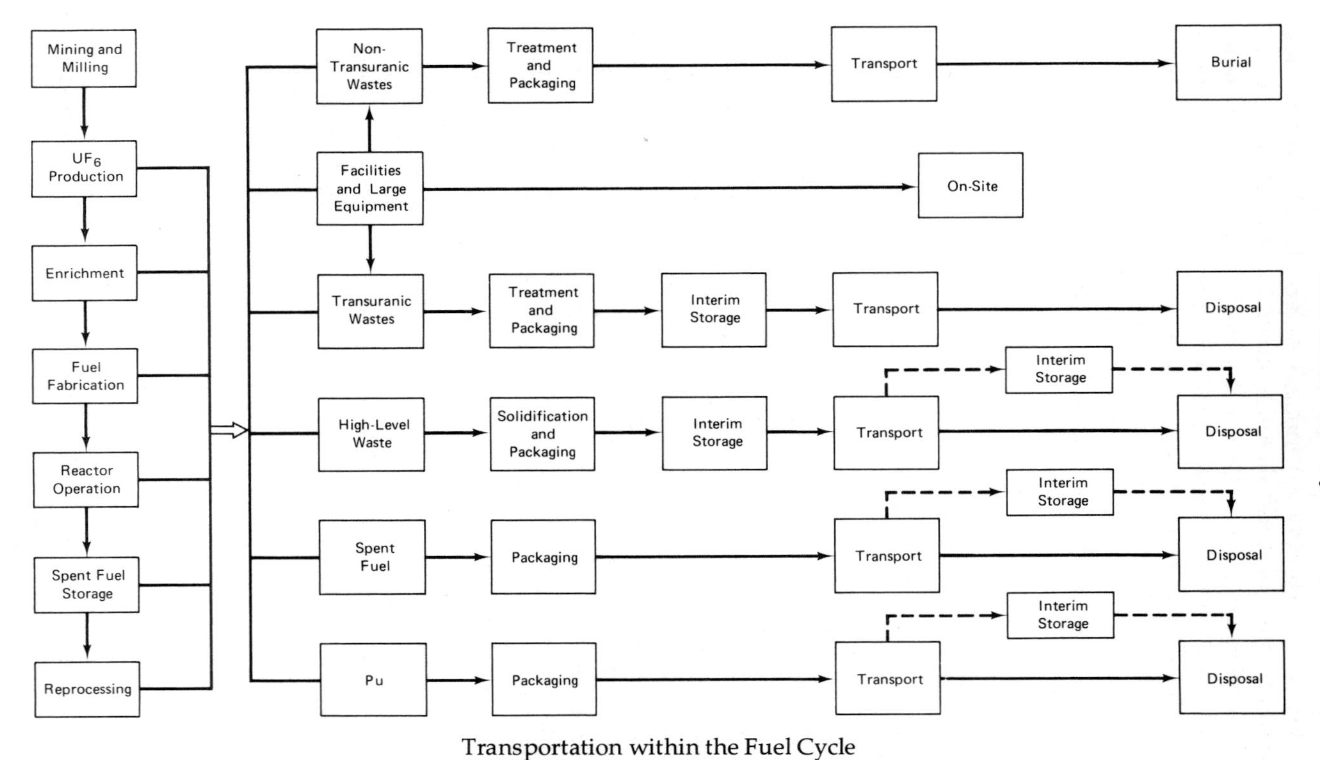

Transportation within the Fuel Cycle

> truck at the port of Savannah, Ga. Jerry Morris, a state radiation safety officer, asked the driver of the truck if he would travel a designated route. The driver said, 'there was no assigned route'. He said further that he had no specific instructions in case of an accident. He did not know what cargo he was carrying. There was no radiation monitoring equipment. The shipment was not properly labeled. Fortunately, no accident occurred en route to the Savannah River Plant, near Aiken, SC.[397]

Given this type of carelessness it comes as no surprise to learn that between 1971 and 1975 there were 36 accidents involving radioactive materials that resulted in the release of excess radiation levels into the atmosphere. These 36 accidents were amongst 144 other accidents.

In the village of Southminster in England on 5 June 1979, a 50-ton container of nuclear waste was left completely unattended for almost an hour, during which time children wandered through the railway yard where it was left to look at it and play around it. Three officials of the Bradwell nuclear reactor finally turned up and directed workers to load the container on a special train bound for the Windscale reprocessing plant.[398]

These nuclear materials are not harmless, but like radioactivity in all the phases of the fuel cycle, can cause extreme environmental and human damage. It has been estimated that if only 1/10th of one of the high-level radioactive fuel containers that are currently being transported through London as a matter of routine was to be released into the environment because of an accident or sabotage, there would have to be an evacuation for several kilometres in each direction, with the possibility of radioactive contamination for at least 100 years.

Nevertheless, for something as toxic as plutonium, the most dangerous radioactive substance known, accidents such as the above are allowed for as the regulatory agencies and the industry plan out the plutonium economy into the 1980s. This can be seen from the chart on page 199 which shows the NRC projection for plutonium shipment in the next decade.[399]

9. Waste Management

The final outcome of the nuclear fuel cycle and the final destination of its transportation system is waste – radioactive waste. Because the radioactivity is so toxic, extreme care must be

NRC-proposed System for Shipping Plutonium in the 1980s

	Transport mode				
	Truck		Rail		Aircraft
Shipping container	6M	L-10	6M	L-10	6M
Containers per shipment*	39	50	90	68	39
Quantity of plutonium per container, kg	2.55	2.0	2.55	2.0	2.55
Average shipping distance, km	2500	2500	2500	2500	2500
Total plutonium shipped, metric tons	18	18	18	18	18
Total number of shipments	180	180	90	144	180
Accidents per kilometre	1.5×10^{-6}	1.5×10^{-6}	8.1×10^{-7}	8.1×10^{-7}	5.9×10^{-9}
Accidents per 1,000 MT	166	166	10.125	12.15	0.15

* The number of containers per shipment is based on efficient packing of the transport vehicle consistent with criticality safety requirements. Plutonium safeguards requirements may necessitate shipment of fewer containers in each transport vehicle.

taken to ensure the protection of the public. The present system attempts to do so through two primary types of waste management: immediate release and isolation and delayed release.

Immediate release into the air or water is allowed at all stages of the nuclear fuel cycle. The limits set are generally, according to the NRC, 'as safe as practicable' or 'as low as reasonably achievable'. When Met-Ed released radioactive effluents into the air and into the Susquehanna river during the Harrisburg crisis it was following this 'low as reasonably achievable' method of waste management. So, too, was Getty Nuclear when it released 250 to 500 pounds of enriched uranium into the river above Jonesboro, Tennessee. If this was not allowed, all nuclear facilities would have to shut down, as such releases are built into the very designs and operation of the plants.

The isolation and delayed release type of waste management attempts to contain all the radioactive waste *not supposed to be released into the environment*. There are two stages in this programme: temporary storage facilities and permanent storage facilities.

The largest temporary storage facility in the world is the Hanford Nuclear Reservation in the south-eastern section of the state of Washington, along the banks of the Columbia river. There, fully 75% of all US radioactive wastes are stored. Concentrated radioactivity so hot that it boils from its own heat is kept in enormous concrete clad steel tanks. Some tanks have been built to hold 1 million gallons of liquid. Many tanks are water cooled, allowing the radioactive liquid to boil off, leaving a solid sludge.

The federal government took over this huge complex in 1943 in order to produce plutonium for the atomic bomb project. Eventually, a nuclear materials processing plant and nine reactors were to be built. The plutonium produced was used in the Nagasaki bomb and then later for the growing stockpile of nuclear weapons. Much of the plutonium was solidified and sent to Rocky Flats, Colorado, Nuclear Weapons Facility to be used in making the plutonium triggers for the American arsenal of hydrogen bombs.

Despite the fact that this aspect of the fuel cycle is for radioactivity that is not to be allowed into the environment at all, the Hanford Reservation has been plagued with leaks, accidents, and operation and design failures. By 1973, many of the tanks were between twenty and thirty years old and nearing the time when they needed to be replaced. This situation was known about inside the AEC but it was reluctant to make the type of financial investment necessary to build new tanks. Consequently, as of 1978, a total of more than 500,000 gallons of radioactive effluents have leaked out of their holding tanks.[400]

One of the spills involved the release of 115,000 gallons of high level waste, releasing 14,000 curies of strontium-90, 40,000 curies of cesium-137 and 4 curies of plutonium directly into the soil. The leak went completely unnoticed for over two months. An investigatory report by the AEC stated that worn-out tanks and primitive monitoring systems accounted for both the leak and the fact that it went unobserved for so long, but that the primary cause of the spill was due to the negligence of the Atlantic Richfield Hanford Company which was contracted to supervise the day-to-day operation of the waste storage at Hanford.

What was *not* noted by the AEC in its report was that *it* was the agency responsible for making sure that Atlantic Richfield did its job correctly. It had known about serious defects in the waste management system from a General Accounting Office report in 1968, but had kept the information classified instead of acting upon it. An even earlier warning from the US Geological Survey team was kept classified until 1973. A third study, too, this one from the National Academy of Sciences in 1966 concluding that AEC waste management practices were seriously unsafe, was classified until 1970, when Senator Frank Church compelled its release for public scrutiny.[401]

This negligence becomes serious when it is realized that the Hanford Reservation is immediately adjacent to the Columbia river and its underlying water table. As early as May 1964, the Federal Water Pollution Control Agency said that phosphorus-32 and zinc-65 were contaminating the fish and oysters in the river. The Agency charged the AEC with a negligence that 'has resulted in the worldwide acknowledgment of the Columbia River as the "most radioactive river in the world" '.[402] The suppressed report brought to light by Senator Church indicated that radioactive isotopes of tritium and ruthenium seeping into the ground might eventually contaminate the Columbia's underground water table.

Radioactive wastes spilled into the Columbia do more than contaminate the northwest part of the US. Radioactive effluents can spread from the mouth of the river throughout the northern reaches of the Pacific Ocean, reaching as far away as Japan.

However, neither the AEC or its successor, the NRC, has been worried about the risk of ocean contamination. In fact, the use of the ocean as a dumping ground for radioactive wastes has long been part of the AEC/NRC method of waste disposal. From 1946 till 1970, the AEC simply dumped the waste on to the ocean floor in concrete cylinders that soon began to leak because of the intense corrosiveness of the salt water. Two particular sites, one off the Farallon Islands forty miles west of San Francisco, Ca, and one 120 miles east of the Maryland-Delaware coast received over 97,000 curies of radiation apiece. In 1976, traces of plutonium were found on the beaches of the Farallon Islands. Several European countries have together dumped over 240,000 curies of radiation into the North Atlantic.[403] From the Windscale plant alone, 300 curies of radiation are dumped into the Irish sea *per month*, in planned and allowed releases.

In the mind of someone like Jacques Cousteau, the great explorer of the oceans, such ocean dumping is an 'unatonable crime'. He emphasizes that our planet can only sustain life because of its water, and that while it may appear to us to be in enormous amounts it is only a thin film covering the surface of our globe. Water on our planet, he says, is comparable to the vapour left by a breath on a twelve-inch sphere. Yet this vapour serves to keep the balance of life in equilibrium, for the oceans are the thermal regulators which keep the planet climatologically stable within a few degrees of temperature world-wide, year-round. The oceans, therefore, provide the Earth with its biosphere, its ability to maintain life. To assault the oceans is to assault the mother of life. The ocean, says Cousteau,

> has the capacity to absorb and neutralize a certain quantity of degradable pollutants every year, but as far as non-degradable pollutants are concerned, they just pile up in the sea and are biologically concentrated in the flesh and in the entrails of marine creatures; as for those indestructible toxic materials . . . the only acceptable discharge level is, in the long run, zero. As a consequence . . . nuclear waste should be isolated forever from the water system.[404]

The problem we are faced with, however, is that after three decades of nuclear weapons production and two decades of nuclear reactor operation we have produced tens of thousands of tons of waste so toxic that mere fractions of an ounce can kill and/or mutate life and which remain toxic for countless thousands of years: for example radon-222, when derived from thorium, has a half life of 80,000 years. Plutonium-239 has a half life of 28,000 years.

We have produced nuclear materials, therefore, that we *have* to isolate. There is no other option. And yet where can we put them? What system can the human mind construct that will contain radioactivity from the outside environment for a longer period of time than human beings have even been civilized? There have been a variety of ideas, of course, ranging from shooting rockets full of radioactive waste directly into the sun, to burying the wastes in the polar ice caps where their heat would allow them to burrow down until they hit bedrock, to burying the waste in shafts drilled in the ocean floor, to burying the waste under the continental shelves, to using atomic bombs to blow underground caverns large enough to dispose of the waste, to simply attempting to find geological formations of salt or granite that have not moved much in the last several thousand years and burying the wastes in them.

As of 1980, no single viable solution has been offered for the disposal of radioactive wastes for as long as they must be completely isolated from the environment. For this reason, all high-level nuclear wastes produced in the US and in Europe still remain in temporary storage facilities. West Germany has buried some low-level waste in the salt caverns of Asse, but in the US things are back to square one. In February 1980, President Carter cancelled the long-touted Waste Isolation Pilot Project in New Mexico that was to have been America's first permanent storage site. Instead, he directed that the agencies involved look into sites near Hanford, Washington, Jackass Flats, Nevàda, and into the salt domes of Louisiana for a permanent burial deposit. The site must be chosen by 1985. In the meantime, the President ordered the construction of several federally-owned temporary waste-disposal facilities.

Most serious experts feel that a permanent waste-disposal site that will remain geologically stable and water free for the tens of thousands of years necessary will be impossible to find, at least on planet earth. What the 1973 Pugwash Conference on World Affairs stated still holds true, that

> with respect to the management of long-lived radioactive wastes, strong uncertainties still exist. The principle difficulty is that the material remains highly toxic for periods measured in thousands of years; even over shorter spans, predictions about the stability and continuity of human society are impossible, and over the longer term significant geological change is possible. . . .[405]

The only ultimately viable option may be to remove high-level radioactive wastes off the planet and shoot the wastes either into the sun or into the inner recesses of our galaxy.

10. Summary

In summarizing the facts presented so far concerning the nuclear fuel cycle, the following points should be emphasized:

1. At each stage of the fuel cycle radioactive effluents are *legally* released into the air and water. This is because of the acceptance of the *threshold theory* which asserts that a human body can sustain 'safe' dosages of radiation. Without the acceptance of this theory, nuclear facilities could not be built, for radiation releases are inevitable.

2. All recent studies indicate that there is no such thing as a safe level of radiation exposure. *Any* radiation is potentially harmful whether naturally occurring or human made. This is why around nuclear facilities of all sorts – from mines, to mills, to conversion plants, to enrichment plants, to fuel fabrication

facilities, to nuclear reactors, to reprocessing plants, to nuclear weapons factories, to waste disposal sites – both the *workers* in these facilities and the *public* living around the facilities are experiencing dramatic increases in cancers, in leukemia, and in genetic defects. There is no such thing as a 'safe' nuclear reactor or facility any more than there is such a thing as a 'safe' nuclear weapon.

3. Because of the high toxicity of the radioactivity produced and because of the extremely long half-lifes of many radioactive agents – strontium-90, 28 years; radon-222, 80,000 years; plutonium-239, 28,000 years – the damage to human life and environment will not stop with our generation. Rather, each nuclear facility, while itself only lasting thirty to forty years, produces radioactivity which will live on long after both the builders of the plant and the plant itself are gone and forgotten. Our generation, therefore, is producing a radioactive legacy that will contaminate the environment and damge human life for thousands of generations to come.

4. There is at present no *place* to store radioactive wastes and no *technology* known which is capable of containing the long-term high-level wastes for as long as they must be completely isolated from the human and natural environment.

5. If the nuclear fuel cycle was stopped, the present loss of life and the prospect of leaving a radioactive legacy to future generations could be largely avoided.

III

Karen Silkwood: A Life in Death

So far I have discussed the psychological and physical dimensions and impact of the nuclear weapons/reactor complex. Nuclear weapons were produced in a climate of war and fear and have resulted in a heightening of paranoia and global mistrust to the point that history has entered the Age of Overkill. The situation has been intensifying in recent years because more and more countries are developing nuclear weapons programmes, largely from the plutonium waste generated by their civil nuclear operations. The atoms for peace programme has increasingly been turned into atoms for war. Even taken on their own terms, however, nuclear reactors, indeed, all aspects of the nuclear fuel cycle, are seen to be as devastating as the nuclear weapons used against Hiroshima and Nagasaki. The workers in the plants and the populations around them have been physically damaged, and thousands of people have died from radiation-induced cancers and leukemia. The only difference between nuclear weapons and nuclear reactors is that one goes off with a huge blast while the other releases its radioactivity slowly, quietly, over time. But the effects are the same: environmental damage and human death and mutation.

What still needs to be discussed is a third dimension of the nuclear power phenomenon. Besides their impact on the human psychology and the human body, nuclear weapons and the plutonium economy have an immediate impact upon human *freedom*. This point can be seen by recalling a fact noted earlier, that each operating nuclear reactor produces between 400 to 600 pounds of plutonium waste each year. When one considers that less that one millionth of a gram, if ingested, is enough to cause cancer and/or genetic mutation, and that twenty pounds, if properly fashioned, can be made into a nuclear bomb, it becomes obvious that *the different aspects of the plutonium economy where either enriched uranium or plutonium are obtainable must be as tightly guarded as nuclear weapons are*. While nuclear weapons are kept at military facilities, however, generally away from population centres and specifically under guard in a military system predicated upon discipline, hierarchy, and authoritarian leadership, protecting the atoms for peace programme, because it was designed for service in the civilian non-military part of society, will have a devastating impact upon the democratic freedoms and civil liberties of the citizens if it must be as closely protected as the nuclear weapons are.

The potential problem with the plutonium economy and its relation to human freedom has been succinctly expressed by a statement made by Dr Bernard Feld, Chairperson of the Atomic and High Energy Physics Department of the Massachusetts Institute of Technology:

> Let me tell you about a nightmare I have. The Mayor of Boston sends for me for an urgent consultation. He has received a note from a terrorist group telling him that they have planted a nuclear bomb somewhere in central Boston. The Mayor has confirmed that 20 pounds of plutonium is missing from Government stocks. He shows me the crude diagram and a set of the terrorists outrageous demands. I know – as one of those who participated in the assembly of the first atomic bomb – that the device would work. Not efficiently, but nevertheless with devastating effect. What should I do? Surrender to blackmail or risk destroying my home town?[406]

The dangers are real, so real that government planners in every country with nuclear programmes have undertaken steps to be prepared for Dr Feld's scenario. In 1975, the NRC commissioned a specific study of the problem. One of the participants, Professor John Barton, Professor of Jurisprudence at Stanford University Law School, prepared a paper entitled 'Intensified Nuclear Safeguards and Civil Liberties'. The document began by stating that

> Increased public concern with nuclear terrorism, coupled with the possibility of greatly increased use of plutonium in civilian power reactors, are leading the US Nuclear Regulatory Commission (NRC) to consider various forms of intensified safeguards against theft or loss of nuclear materials and against *sabotage*. The intensified safeguards could

> include expansion of personnel clearance programs, a nationwide guard force, *greater surveillance of dissenting political groups*, area searches in the event of a loss of materials, and creation of *new barriers of secrecy* around parts of the nuclear program.[407]

It is important to be clear what the above statement implies. The governments supporting nuclear power are attempting to protect the plutonium economy from two perceived enemies: first, those who would use the nuclear materials to terrorize the country through some type of nuclear sabotage; and second, those who seek to stop nuclear power, meaning 'dissenting political groups' who have anti-nuclear stands. This requires a nationwide guard force to be created specifically to deal with any terrorism and the erection of new barriers of secrecy around the nuclear programmes to keep public knowledge and participation at a minimum. Both sets of enemies would be subject to greater surveillance through electronic listening devices such as phone taps.

The country that has gone the farthest in officially eroding the civil liberties of its citizens in the interests of protecting the plutonium economy has been Britain. While in other countries any political and character screening of workers in nuclear installations is kept discreet and officially denied, in Britain it is accepted as a matter of course that anyone working for the Atomic Energy Authority be 'positively vetted' before being appointed. The Official Secrets Act, moreover, allows the government and the atomic industry to keep the nuclear installations cloaked in secrecy and the employees forbidden to communicate anything about their work. In 1976, Britain also became the first country to establish by law a nationwide guard force of constables under the direct control of the atomic authorities in order to guard nuclear facilities and specifically the plutonium kept there. This guard force has privileges in relation to carrying weapons not granted to any other British police unit. Indeed, so sensitive are these privileges that under the Official Secrets Act they have not even been made available to public knowledge. This force is mandated not only to guard against possible terrorism but to keep tabs on 'dissenting political groups'. What this means, according to Michael Flood and Robin Grove-White, of Britain's Friends of the Earth organization, is that

> Anyone who has ever expressed himself critically on questions of nuclear energy must count on such attempts at surveillance. Fear of 'civil disobedience' might be sufficient to cause the security services to make a careful check of suspects, including their private lives. The already existing force of special constables responsible for atomic security will in the future have an establishment of at least 5,000 men. On top of this there will be a large number of guards all equipped with the latest technical equipment for 'data finding' – a euphemism for espionage that in this case is almost exclusively directed at fellow-nationals.[408]

Jonathan Rosenhead, of the London School of Economics, points out that this type of political control is very easily unnoticed by the general populace because it is specifically designed and intended to be used as inconspicuously as possible. In America, political scientists refer to this technique as the 'politics of the iron fist in the velvet glove'. 'What the ruling groups prefer', he says,

> is to produce a situation in which no one dares oppose their plans. Their favourite methods are therefore to exploit people's dependence on consumer goods and on their jobs and exercising prevention controls by means of intensive surveillance. In the event of open conflict breaking out in spite of that, they would hope at least to contain it by 'limited operations'.[409]

While Britain has perhaps gone the farthest in *officially* eroding the civil liberties of its citizens in order to protect the plutonium economy, the country that has perhaps done the most systemic *unofficial* damage to human rights has been the United States, where under a veneer of democratic rights an extensive police-state network is being constructed which includes not only legitimate police units but industry and utility company security groups as well. Rather than give data on these groups, however, it might be more helpful to examine the plutonium economy and its impact upon civil liberties as it touched one solitary life: the woman Karen Silkwood. Like the *hibakusha* of Hiroshima and Nagasaki, her life and interaction with the nuclear establishment personifies what the plutonium economy implies for humanity in general.[410]

In early 1972, Karen Silkwood, a 26-year-old mother of three children, went to work for the Kerr-McGee (KM) Nuclear Facility in a little Oklahoma town named Cimarron. Trained in physics and a firm believer in the benefits of nuclear power, she was hired as a laboratory technician to test the quality of the plutonium fuel being produced at the plant. KM had been awarded a multi-million dollar contract to produce the plutonium fuel rods for an experimental plutonium-fired Fast-Flux Breeder Reactor being built by the Westinghouse Corporation near Richmond, Washington.

Once working, Karen became quickly involved in the efforts of the Oil, Chemical and Atomic Workers International Union (OCAW) to organize the workers into a local branch of the OCAW. The main reason for organizing was worker complaints about the lack of regard for their health and safety. Many were put on the plutonium production lines without any training at all; the physical plant leaked highly radioactive materials out on to the floor, contaminating dozens of workers at a time; and KM management, in order to meet production quotas, ordered workers to stand in contaminated areas and to continue working while clean-up crews attempted to decontaminate the area around them.

By the beginning of 1973, the situation had deteriorated to the point where the OCAW, which had by then won union status, brought the workers out on strike. Non-union workers were brought in, however, and, completely untrained, were put to work on the plutonium production lines. KM was going to prove that no union could 'break' the tough oil-men of the KM Corporation.

KM proved too big to beat, and by the late spring of 1973, the union was forced to return to work under a weaker contract than the one they had before the strike. Seizing this initiative, during the summer of 1973 KM began a campaign to 'de-certify' the OCAW as the union authorized to represent the workers in the 1974 re-negotiation of their contract. As part of this de-certification campaign, KM issued formal written orders forbidding the workers to discuss working conditions with anyone outside the plant, claiming that such discussions 'compromised the security' of the nuclear installation. Workers were also specifically forbidden to talk to anyone from the news media. When some uranium pellets were found scattered around the plant, KM management seized upon the opportunity to force workers to undergo lie-detector tests. They were subjected to questions about their union support; their contacts with union organizers; their sex lives; their use of drugs; and whether or not they had ever made any 'anti-Kerr-McGee' remarks to anyone from the news media.

Those who refused to take these tests were either fired or transferred to the wet ceramic divison of the plant, an area with an extremely high incidence of radioactive contamination, and were reported to the Federal Bureau of Investigation (FBI) as 'suspects' in the uranium pellet scattering incident.

In July 1974, the Corporation forced the 'de-certification' election – and narrowly lost. OCAW thus was assured of the right to represent the workers in the new contract negotiations scheduled to begin on 13 November 1974. In August, Karen Silkwood (along with Jack Tice and Gerald Brewer) was elected to be a union negotiator. By December of 1974, Brewer had been fired, Tice had been transferred to the wet ceramic area, and Silkwood was dead.

The following is an account of the events leading up to her death.

When Silkwood was elected to be one of the union representatives, her area of responsibility was the issue of health and safety. The union had been crushed in its first strike over the health and safety issue back in 1972, and she was determined not to have it happen again. She interviewed all the workers who had ever reported KM violations of the health and safety rules; she memorized the Atomic Energy Commission (AEC) regulations promulgated to safeguard the health of the workers; and she conducted her own health and safety inspections of the plant during her free time, compiling a long list of violations she herself noticed.

Knowing what the health and safety regulations were and seeing the violations of these rules by KM, Silkwood flew back to Washington, DC, in September 1974 and testified behind closed doors before the AEC, asserting that KM was guilty of violating some forty different health and safety regulations. The response of the AEC was to tell Silkwood and the OCAW that they would have to further 'document' these accusations before it would act in any way.

On 27 September 1974, during her stay in Washington, Silkwood first informed the OCAW that KM was not only guilty of gross violations of health and safety regulations but was seriously violating quality control standards as well. She told Tony

Mozzochi, Vice President of the Union, and Steve Wodka, Mozzochi's assistant, that she could get documented proof that KM officials were knowingly 'doctoring', with *magic marker*, the safety inspection X-rays which, by AEC regulation, had to be taken of each plutonium fuel rod to insure that it was not leaking radiation through faulty welding. Tony Mozzochi was later to testify in federal court that these charges by Silkwood 'were the most serious charges I have heard in my trade union experience'. The AEC demanded perfectly welded fuel rods because, if defective, they could cause a disturbance in the flow of the liquid sodium used to cool the reactor core of the fast breeder. If the sodium is blocked from reaching the portion of the fuel it is meant to cool, the fuel rods can overheat, leading to accidental releases of radioactivity. Under worst-case conditions, faulty fuel rods can lead to a meltdown of the reactor itself.

Karen was instructed by Mozzochi and Wodka to return to her job in Oklahoma and secure copies of the 'doctored' X-rays along with as much documentation as possible of the forty other charges she had made of KM health and safety violations.

Karen returned to Oklahoma on 1 October 1974, and spent the entire month working at the plant by day and by night surreptitiously entering KM executive offices and laboratory facilities in order to photocopy internal KM documents proving her accusations made before the AEC. By the end of October, she had conclusive documentary proof of the validity of twenty-five of her forty charges. On 1 November, she finally secured copies of two, separate, 'magic-marker-doctored' fuel rod safety inspection X-rays, doctored by one Scott Dotter, the special laboratory technician who the KM executives had specifically assigned to conduct the final safety inspection X-raying of the fuel rods. Silkwood discovered that not only was he doctoring up the X-rays indicating faulty welds, but that the mere *number* of the fuel rods he 'cleared' each week was *itself* a direct violation of AEC regulations requiring that no one inspector be allowed to give the final clearance on over a certain percentage of the fuel rods leaving the plant.

On 2 November Karen telephoned Wodka in Washington and informed him that she had secured the documentary proof for twenty-five of the forty charges against KM and that she 'had gotten the goods on Kerr-McGee': on the 'other matter' which they had discussed. Since the union negotiations were coming up on the 13th, Wodka advised Silkwood to 'lay low' until a few days before the negotiations were to start. At this time he would fly out to Oklahoma with Dave Burnham, an investigative reporter of the *New York Times*. Karen would then turn over to Wodka and Burnham all of her documentary evidence sustaining her charges, including her charge of wilful falsification of quality control reports. The plan was that Burnham would write up these charges and print them in the *New York Times* during the first week of the contract negotiations.

Unknown to Silkwood and Wodka, KM executive officials were aware of her every move. Nearly a month before, on 12 October, Oklahoma City Police Department Intelligence Unit Photographer Bill Byler and his friend and unofficial assistant, Steven Campbell, made contact with Silkwood while she and a friend, Drew Stevens, were at a restaurant. They 'discussed' with Silkwood her status as an employee of the KM Nuclear Facility; her alleged 'anti-nuclear' attitudes expressed in the form of her voiced concerns concerning KM violations of worker health and safety regulations; and her activities as one of the elected OCAW negotiators for the forthcoming contract negotiations. On 14 October, Byler discussed the information he had gained from Silkwood with the Oklahoma City Police Department Intelligence Unit Commander Bob Hicks. Hicks told Byler to gather as much additional information about Silkwood and her associates as he could and to report that information to the Intelligence Unit Commander as he obtained it. Hicks also gave to Byler the private unlisted number of the Director of Security for the KM Corporation, James Reading, instructing Byler to contact Reading and convey to him all the information he had gathered on Silkwood. On 15 October, Byler telephoned Reading and met with him on the same day. Both Reading and Hicks were kept informed of all subsequent information gathered on Silkwood.

The next move Byler and Campbell made with regard to Silkwood was to photograph pages of a confidential diary Drew Stevens was keeping in which he was cataloguing the activities of Karen and her efforts to stop the violations of health and safety standards at the plant where they both worked. Copies of the photographs made were given to both Hicks and Reading. Within this same time-frame Karen's phone was tapped and all her telephone conversations monitored. Moreover, Byler, Campbell, Hicks and Reading began to meet privately to discuss the possibility of finding some possible criminal charges against

Silkwood and her associates – possibly relating to the paraphernalia Byler saw in Karen's apartment which he suspected was used to smoke marijuana.

After the 2 November telephone conversation with Wodka in Washington, Karen took two days off from work. She returned to work on 5 November and performed her usual duties as a lab technician. This included working behind a glass shield, through glove-box openings and rubber gloves, to test various samples of fuel batches.

When she stopped to take her coffee break, she undertook the standard safety check of her person before leaving the glove-box area. She registered radioactive contamination. She immediately called for help and was taken to the decontamination area where she was undressed and scrubbed with wire brushes and treated with chemicals to remove the plutonium from her skin.

The safety inspection team then inspected her glove box but could find no radiation escaping from it; nor was there any radioactive contamination in any part of the room she had been working in. The KM Corporation was later to admit in depositions that the contamination on Silkwood did not come from that room. Nor did she receive the dosage from any other part of the plant.

After the long scrubbing and decontamination process, Karen went home. She returned to work the next day and again when she began her check-out from the glove-box area she registered radioactive contamination. She was again rushed to the decontamination area and put through the process of attempting to remove the plutonium from her skin. This time, however, she registered internal contamination of her nasal passages. She was ordered to go home and to return the first thing in the morning to undergo a 'sinus draining' procedure designed to lessen the continuing intake of plutonium into her lungs – the part of the body most sensitive to the carcinogenic effect of plutonium.

Once again the laboratory area in which she had been working was inspected; again it was entirely free from radiation with the exception of a few places which she had touched with her contaminated hands.

On the morning of 7 November Karen reported directly to the decontamination area of the plutonium facility for the 'sinus draining' procedure. Upon her first inspection, she was found to be heavily contaminated around her face, neck, shoulders, arms and hands. She had not been to the laboratory, she had not been in any other part of the plant; in fact the only place she had been since she had been decontaminated the day before was her home.

KM ordered Silkwood to fly immediately to Los Alamos, New Mexico, to undergo a full week of special physical tests to determine the extent of her internal contamination. It was indeed unfortunate, they informed her, that she would have to miss the contract negotiations she had been preparing for but health and safety must come first.

On 8 November, Karen was sent to Los Alamos, to a special US government radiation laboratory where she was to be put through a week-long battery of tests and examinations.

Once Silkwood was gone, a Special Inspection Team from the KM Nuclear Facility descended upon her home to begin what one of the team later described in sworn testimony as 'a full-scale search'. Karen's personal mail was read and all tape recordings, personal diaries, notes, memoranda and other documents having anything to do with either her work at KM or her alleged use of marijuana were turned over to James Reading, the Chief of KM security, *not* to the decontamination team.

During the search, the KM Special Investigation Team was joined by federal inspectors from the Region III office of the AEC. On 8 November, the AEC inspectors opened up Karen's kitchen refrigerator and found that the bologna and cheese she had been eating was radically contaminated with 400,000 disintegrations per minute of plutonium radiation. The apartment was immediately sealed off, and *all* of its contents taken, leaving only the bare concrete walls and floors. The contents were sealed up in 55 gallon drums of the type used by KM to dispose of its plutonium waste. These barrels and their contents have never been seen again by anyone other than KM and AEC officials.

On the night of 12 November, although Karen was scheduled for more tests, she gathered her belongings and flew from Los Alamos back to Oklahoma. She telephoned Wodka and told him that despite the loss of everything in her apartment she 'still had the documents' and that as far as she was concerned the plan to meet Wodka and Burnham the next evening was 'still on'.

Karen spent the night of 12 November in Oklahoma City, showing up at the KM facility, some 40 miles away, at 9:00 a.m. on the morning of 13 November for the contract negotiations.

She negotiated with Tice and Brewer during the day, not mentioning the documentary evidence in health and safety and wilful violation of quality-control regulations.

After the negotiations broke off in the late afternoon, Karen went with other union members to the Hub cafe in Cresant for a de-briefing session. She took one short break from this meeting to telephone Drew Stevens to make sure he was leaving his work early to go to the Oklahoma City airport to meet the planes on which Burnham and Wodka were arriving, to make sure they got to the 8:00 p.m. meeting on time. After this call, Karen went back to the de-briefing and confirmed to a friend, Jean Jung, that she had documentation concerning health and safety and quality control violations that would 'get Kerr-McGee once and for all'. She also told Jean Jung about the meeting she was to shortly leave for with Wodka and Burnham.

At about 7:10 p.m., Karen left the cafe, got into her 1973 Honda Civic with an inch-thick manila folder full of documents, and drove off down Route 74 toward Oklahoma City.

In Oklahoma City, Burnham, Wodka and Stephens sat waiting for Karen in Burnham's hotel room at the Holiday Inn Northwest. Eight o'clock came and went. No Karen.

Eight-thirty came and went. No Karen.

The men then began to become concerned and attempted to call the Hub cafe. They discovered, however, that Burnham's phone was 'out of order'. Leaving the room, they finally made contact at a pay phone and learned that Karen had left the cafe shortly after 7:00 p.m.

The men jumped in their car and headed out of Oklahoma City along Route 74 towards Cresant.

Seven miles from Cresant, they came upon the red lights of police cars and a gathering of spectators. Karen was dead. Her body was gone. Her car had been taken away.

Telephone calls from David Burnham to the Oklahoma State Highway Patrol revealed that Karen's car had been towed away by some private wrecking service, *not* by the State Highway Patrol, as was the usual procedure. An all-night search by Burnham, Wodka and Stevens failed to turn up where Karen's car and her documents had been taken.

Before searching for the car, the men went to the coroner's, viewed Karen's mangled body, and telephoned her parents in Nederland, Texas, telling them of their daughter's death.

The car had in fact been taken away from the accident by a

local wrecker service, Ted Seabring's Ford. At 12:05 midnight, he was called by the Oklahoma State Highway Patrol and directed to open up his garage in order for a group of KM officials to check Silkwood's car, 'to inspect it for radiation'. Four KM representatives came, dressed in radiation suits, complete with face masks. Both Ted Seabring and his assistant Kenneth Valliquet later asserted that these men took documents from the car and began reading them aloud to one another. Neither Seabring nor Valliquet recall whether these men actually took the documents they were reading.

Later that morning, police officials came to the garage and did take items from the car. After this visit, Seabring took all the remaining items and packed them into a cardboard box, which he sealed.

It was not until that afternoon, 14 November, that Burnham, Wodka and Stevens finally located Karen's car. After a call to Karen's parents, Seabring released Karen's car and the sealed box of items he had taken from the car to them. Directly in the presence of Seabring, Burnham, Wodka and Stevens opened the sealed box and examined its contents. There was no manilla folder. There were no documents.

Wodka immediately telephoned Tony Mozzochi, OCAW Vice President. The Union decided immediately to hire the best automobile crash reconstruction laboratory in the Oklahoma area to inspect the car and the crash scene in order to reconstruct what had happened. Contracted for the job was the Accident Reconstruction Laboratory in Dallas, Texas. One of their top experts, A. O. Pipkin, flew to Oklahoma, inspected and photographed Karen's car; inspected and photographed the accident scene; and reviewed both the police reports and the report of Seabring. On 19 November, the Accident Reconstruction Laboratory issued an official report concluding that the physical evidence Pipkin had been able to evaluate indicated that Silkwood had been struck in the left rear side by another automobile travelling up behind her at approximately 55 to 60 miles per hour. She had been knocked off the road and her car had driven directly into a concrete culvert. Karen had been killed instantly.

On 20 November, the OCAW filed a formal complaint with the Justice Department in Washington, DC, demanding that an investigation be conducted into Karen's mysterious death as well as into her mysterious contamination a week before her death. On 21 November, FBI Headquarters in Washington telexed the FBI office in Oklahoma City and ordered a full-scale investigation into the contamination and death of Karen Silkwood and into charges of union harrassment by KM. FBI agent Lawrence J. Olsen was assigned to the case.

Sworn depositions reveal that on the day he was assigned to the case, Olsen met privately with James Reading, informing him that the FBI had ordered him to investigate KM agents for possible criminal wrong-doing against Silkwood and other members of the OCAW at their Cimarron Nuclear Facility. At this meeting, Olsen explained to the chief of KM security what type of information might prove incriminating to KM and asked Reading to set up an emergency meeting between Olsen and executive officials of KM as soon as possible.

This second secret meeting took place on 25 November. After explaining to the KM officials what the federal statutes were under which he had been ordered by FBI Washington to conduct a criminal investigation against KM, Olsen agreed to work with Reading in the investigation. During the following few days, Olsen interviewed several OCAW workers at the KM plant in a meeting set up by the Company and personally attended by Reading. At these meetings, the workers were asked whether they had any information they wanted to give concerning KM harrassment of the OCAW. When none of the workers came up with any incriminating evidence, Olsen closed his investigation of charges of KM harrassment of OCAW union workers.

Olsen then turned to investigating Silkwood's radioactive contamination. Instead of investigating the possibility of KM contaminating her, however, a confidential memorandum sent by KM Security Chief Reading directly to Dean A. McGee, Chairperson of the Board of Directors of the KM Corporation, indicates that the 'thrust' of Olsen's investigation was to discover 'what were the means used *by Silkwood* to remove plutonium from the Cimarron Facility, or if in fact, the plutonium she was contaminated with came from some other source'.

In March 1975, Olsen met with an undercover FBI domestic surveillance operative Jacque Srouji who had been asked by the FBI to write a book on nuclear power in order to 'establish contacts in this area'. Olsen turned over his entire Silkwood investigative file to Srouji for copying. He also set up a secret meeting between Srouji and Reading. Srouji was later to state to a Congressional investigator that at this meeting Reading informed her that he and his associates in the Security Divison of KM had wiretapped Silkwood's phone and had electronically 'bugged' her home prior to her contamination and death. Srouji states that Reading even showed her typed transcripts of the telephone calls monitored.

In the late Spring 1975, Olsen closed his investigation, stating that it was not possible to determine how the 400,000 disintegrations per minute of radioactive plutonium got on the food in Silkwood's apartment. In his report on her death, he accepted without question a report issued by the Oklahoma State Highway Patrol stating that Karen had fallen asleep at the wheel. With regard to the question of documents, Olsen stated in his report that KM officials had said that there could not have been any documents since her charges against the company had not been true. This being the case, he never looked for any.

Within a few months of her death, therefore, the FBI investigation of Karen Silkwood was closed. In response to official inquiries by her parents, the National Organization for Women, and the OCAW, FBI Inspector Al Connally responded that they 'watched too much television' if they expected every case to be solved completely. 'In any event', Connally said, 'the FBI cannot discuss its investigations into the case, because it is a "closed" case. And the FBI official policy is not to discuss "closed" cases.'

Unsatisfied with this explanation, the National Organization for Women, the OCAW, the Environmental Policy Centre, Critical Mass, and various other interested organizations began to mobilize their constituencies to put pressure on the Senate Committee on Government Operations to hold a public enquiry into what had really happened. In November 1975, Senator Ribicoff, who chaired the Committee, agreed to hold hearings to find out not only what had happened but to determine how effectively the various federal agencies, including the FBI, had performed their duties.

As soon as the Senate hearings were announced, the KM Nuclear Facility was completely closed down and all of its workers summarily fired. The plant was put into 'mothballs', to await the pending investigation. Suddenly, two weeks before the investigations were to begin, Dean McGee flew to Washington and met privately with Senator Metcalf, a high-ranking member of the Committee on Government Operations who with Ribicoff was heading the inquiry. The only other person present during this meeting was a Mr Bennet, the Washington representative of KM. No one knows what was discussed at the meeting. The results, however, were dramatic. The very next day, Senator Metcalf issued a press release announcing that the previously scheduled hearings into the Silkwood matter were 'permanently closed'.

One of the Congressional investigators for the now-closed Senate hearings, Peter Stockton, took the case over to the House of Representatives and got the House Sub-Committee on Energy and the Environment to agree to hold hearings into the Silkwood matter in April 1976.

When the FBI was contacted to provide copies of its Silkwood investigation file, it refused, citing its official policy not to discuss 'closed' cases. The Chairperson in charge of the enquiry, John Dingell, rejoindered by pointing to cases where Congress had been explicitly authorized to obtain copies of 'closed' FBI investigations, when such files were deemed necessary by the Congress for it to perform its responsibilities. Upon receipt of this information, the FBI immediately declared the Silkwood case 'open', refusing to give to Dingell access to its files on the grounds that the FBI had the right to refuse *anyone* the files on an 'on-going' FBI investigation. When questioned by Dingell how the status of the Silkwood case had suddenly switched from 'closed' to 'open' the Deputy Director of the FBI, James Adams, stated that the case had been re-opened 'due to all the inquiries being directed to the Bureau about it'.

Dingell held hearings anyway. One of the witnesses was Jacque Srouji who, according to FBI documents later obtained through subpoena power, was sent to testify in order to blast the OCAW and defame Karen Silkwood. In her testimony, she accused Silkwood of being 'mentally unstable', of 'deserting her husband and three children', and of 'using marijuana'. She further stated that in her opinion both the contamination and the death of Silkwood had been deliberate and hinted darkly that perhaps the group responsible was the OCAW, which could have done it somehow to embarrass KM in the contract negotiations. Under cross-examination, Srouji defended her conclusions by stating that they were based on FBI documents. This brought the proceedings to a halt, for if Srouji, introduced to the investigation as a journalist, had seen FBI files, why had these files been refused to the Congress of the United States?

Before the issue could be resolved, Dingell was ousted as the Chairperson of the Committee in a *coup d'état* engineered by another member of Congress, Tom Steed, from the fifth District of Oklahoma, the home district of the KM Corporation's international headquarters.

With this, hearings ceased, and Karen's parents, until this time waiting patiently for American justice to deal fairly with the contamination and death of their daughter, contacted legal counsel to file a federal lawsuit designed to obtain justice through the federal courts.

In November 1976, a three count suit was filed charging first, that KM was legally liable for the plutonium contamination of Silkwood which occurred on 5, 6, and 7 November 1974; secondly, that James Reading, Dean McGee, Reading's assistants, and the other members of the KM Board of Directors participated in a wilful and intentional conspiracy to violate the civil rights of Karen Silkwood in her efforts to organize a lawful trade union at KM; and that they then sought to cover up these violations. Four FBI agents, including Olsen and Srouji, are named as co-conspirators in the cover-up. The third count charges these same people with an identical conspiracy of attempting to commit and then cover-up a deprivation of the equal protection of the laws and the right to the equal enjoyment of the privileges and immunities of all those persons, of whom Silkwood was one, who reported violations of the federal Atomic Energy Act by the KM Corporation.

The first count finally came to trial in the Oklahoma Federal Court in the spring of 1979. Dr John Gofman, known to many as the 'Father of Plutonium', because he was one of its co-discoverers, set the tone for the plaintiff's case by testifying that current government licences to operate nuclear plants conforming to existing standards are 'legalized permits to murder'. Evidence from the past ten years, he said, shows that federal standards for plutonium are at least 480 times too lenient and that Silkwood was 'married to lung cancer' as a result of her contamination.

Dr Edward Martell, an environmental radiochemist, also testified that existing radiation exposure limits are 'misleading and inadequate and have not been reduced because of the government's vested interest' in nuclear power. Dr Martell called federal standards both in the US and abroad 'meaningless' because, contrary to official policy, there is no safe limit for exposure to low-level ionizing radiation.

Dr Karl Morgan, often referred to as the 'father of health physics' for his role in the setting of acceptable standards for radiation releases in nuclear facilities, testified that KM showed a 'callous' attitude towards the safety of its workers. He pointed out that the KM training manuals made no mention of the fact that one could contract cancer from radiation exposure, and that because of this refusal to recognize the dangers of radiation both KM management and the workers themselves were lax and in frequent violation of safety regulations.

Former plant workers stated under oath that their training had been so deficient that teenage workers often played at seeing who could get 'the hottest the fastest'. Workers said that plutonium spills were often painted over instead of cleaned up, workers left the plant contaminated, and plant supervisors were warned ahead of time of 'surprise' inspections by the AEC. There was also testimony stating that workers used uranium for paperweights, threw it around the rooms at each other, and even took uranium home to show their children.

One of the four plant supervisors, Jim Smith, branded the KM Nuclear Facility a 'pigpen', testifying that security was so lax, workers could have thrown plutonium over the fence or taken it past guards simply by telling them it was to be thrown out as waste. Smith also told of numerous incidents where workers were forced to stay in contaminated areas and continue working in order to meet production quotas, their only protection being inadequate face respirators.

The jury was convinced by the combined testimony of expert witnesses and former workers, and on 18 May 1979 awarded $10.5 million in actual damages and $500,000 in personal injury damages to Silkwood's three children. In charging the jury, Federal Judge Frank G. Theis directed them to define 'physical injury' with regard to plutonium as 'nonvisible or non-detectable injury . . . to bone, tissue or cells'.[411] The implications of this are profound, for it establishes *legally* that plutonium is in fact a 'dangerous material' and causes 'physical injury'. This means, on the one hand, that nuclear materials are so dangerous that nuclear facilities are under special restraint to prevent the escape of any of the material, whether intentionally or otherwise; on the other hand, it means that workers and members of the public are now entitled to claim damages due to the operation of nuclear facilities if they can demonstrate that their sickness is attributable to radioactive releases coming from the plant involved. In charging the jury, therefore, Judge Theis stated that they did not have to find that KM deliberately contaminated Silkwood; the mere fact that plutonium had been allowed to 'escape' the plant was sufficient to award damages.[412]

In terms of the civil liberties violations of the case, the situation is somewhat more complex. Judge Theis refused to try these counts as he stated that as a union person, Silkwood was not protected against wiretapping and electronic surveillance under the Civil Rights Act. This ruling is under appeal.

What has emerged in the course of the investigation is the fact that the abridgment of Silkwood's civil liberties was not something unique with her. Intentional harassment and illegal electronic surveillance, illegal entry into the home in search of documents, and active co-operation of the FBI in not only the illegal acts themselves but in any cover-up required, are things that many anti-nuclear activists across the US have been forced to undergo. What singles out Karen Silkwood is the magnitude of what she uncovered about Kerr McGee malpractice in the areas of worker health and safety and the extraordinary measures Kerr McGee took to try to silence her. It seems clear that both her plutonium contamination and her fatal accident were deliberate attacks on her.

What needs to be remembered in assessing this state of affairs is that plutonium, if it is to be used, must be protected by police state methods. You cannot have something that can be used for nuclear bombs and can damage and mutate human life with the lethalness of millions of cancer doses per pound in a free society. *A plutonium economy and a free democracy are a contradiction in terms*. This is a fact that has been recognized by leading legal experts and politicians alike. Writing in the *Harvard Law Review*, Russell Ayres stated flatly that 'plutonium provides the first rational justification for widespread intelligence gathering against the civilian population'. The reason for this is that the threat of nuclear terrorism justifies such encroachments on civil liberties for 'national security' reasons. It is inevitable, therefore, Ayres says, that 'plutonium use would create pressures for infiltration into civic, political, environmental and professional groups to a far greater extent than previously encountered and with a greater impact on speech and associated rights'.[413] Sir Brian Flowers, in Britain, has come to similar conclusions. At the conclusion of his environmental impact statement for the plutonium economy in the United Kingdom, known as the Flowers Report, he made it quite clear that Britain could not have both plutonium and civil liberties. Rather, he said, to adopt the plutonium economy would make 'inevitable' the erosion of the freedoms that British people had fought for over the centuries and come to assume as inalienable.

What happened to Karen Silkwood, therefore, is something that should not be seen as an abnormal violation of her person and her freedom; rather, it should be viewed as the *logical conclusion* of what the adoption of the plutonium economy in any country implies. Even as there are certain psychological implications inherent in the use and development of nuclear weapons, then, and even as there are direct physical results on both workers and public alike from the nuclear fuel cycle, so, too, the plutonium economy makes inevitable the erosion of the human rights in the society adopting such an economy.

Again, it should be noted that these erosions are thus far largely subtle and unnoticed, largely because there has not yet been a large scale terrorist incident. But, as Robert Jungk reminds us,

> The question arises when the more severe restrictions on liberty that are expected from an increasing number of atomic institutions will exceed the limits of people's patience and capacity for adaptation. Safety analysts and risk assessors devote a great deal of attention to possible technical 'incidents' and 'blow-ups', but obviously underrate the social 'explosions' that seem to be almost inevitable as a side-effect of their efforts. Meanwhile the very numerous studies of MCA (maximum credible accidents) neglect the likelihood of MCSA (maximum credible societal accidents) which increase from day to day generated by the intolerable pressures of the 'nuclear state'.[414]

What happened to Karen Silkwood was a maximum credible societal accident. If the plutonium economy is allowed to continue there will be others.

IV

Our Challenge: Overcoming Psychic Numbing

In contemplating the nuclear power issue, in terms of both weapons and reactors, one can easily become overwhelmed by the immensity of their potential for damage; by the extent of their permeation into society; and by the enormous effort that would be required of each of us to stop their continued use and development. We resist the problem because it appears too big. T. S. Eliot's remark is all too true, that 'Humankind cannot bear very much reality'.

This resistance is to a large extent natural. It is a way our bodies have of protecting ourselves against too much stress. We are enormously vulnerable to destructiveness, both psychologically and physically. We are equally capable of beauty and creativity. Indeed, both this vulnerability and this potential interact to produce our lives. Both are important in attempting to discern the impact of nuclear power on human beings.

What emerges from the above story of nuclear power is that it assaults our psyches; it damages our bodies; it erodes our freedoms. It is something antithetical to our very humanness. Scientific equations, therefore, cannot be our referent for decision-making on this issue. We cannot say that because something is possible we should do it, and then call this new activity 'progress', regardless of how it affects us. Neither can economic cost-benefit analysis methods be the referent for deciding about nuclear power. Economics involves numbers, currencies, cash-flows – arbitrary units constructed by competitive human minds. Nor can projected ratios of fuel consumption be the final criterion for nuclear power. These resources are of the earth, and nature has her own means of self-protection and self-perpetuation. She cannot be exhausted and even today offers us resources we have not even tapped.

The only referent that can be used is that of the human being. We cannot bow down to science or interlock with an economic system or presume upon Mother Earth. Using these external referents for something that affects us internally is ultimately idolatrous and deceiving. We must examine the price *we* pay for any given enterprise, for any given power we try to shape to make our existence more comfortable and more meaningful. This is why throughout this book the ultimate method of interpreting the nuclear experience has been in terms of its effects upon our psyches, our bodies, our freedoms. We are the measure of our own worth.

Paradoxically, however, this is the very reason for our resistance to dealing effectively with the nuclear weapons/reactor complex: we are forced by them to confront ourselves. This is difficult because we are conditioned to look externally from ourselves for the answers we seek – to governments, to religious dogma, to scientific 'truth', to economic systems. We are taught to look away from ourselves, to give ourselves value only as we interlock with systems outside us. The result is a demeaning of individual worth, a lowering of value in our own humanity – to the point where human life becomes expendable if the 'system' can go on. The La Hague reprocessing plant is a stark example. The workers, forced by their jobs working around plutonium to sacrifice their health, their freedom of speech, their right to self determination, refer to themselves as 'radiation fodder'. 'Realizing that no-one cares,' says Jungk, 'they all fear that after a few years' more work they will end up as "waste" to be dumped on the unemployment rubbish-heap – or, worse still, that they will end up in hospital.'[415]

The result of this dehumanization is psychic numbing, an emotional de-sensitization and acceptance of a lowering of self-worth. Unable to cope, people begin simply to stop feeling. This is a universal phenomenon, one that operates particularly strongly towards anything that threatens our beings with death or damage. It is true that experiences in concentration camps and of the *hibakusha* of Hiroshima and Nagasaki found a certain amount of psychic numbing necessary to deal with the magnitude of the death imagery that had to be faced. But what was discovered was that a little psychic numbing goes a long way, and that too much would cause considerably more damage to the person involved than too little.

This psychic numbing in the face of the technology of the late twentieth century and the all-encompassing bureaucracies interlocked with this technology only causes both to become even more impersonal and dehumanized than they already are. This is particularly true in dealing with the ultimate technological feat of our age: the construction of nuclear weapons of mass destruction and nuclear reactors to power the electricity grids.

Recall the bombing of Hiroshima. The magnitude of technological achievement was such that, according to Arthur Koestler,

> If I were asked to name the most important date in the history of the human race, I would answer without hesitation, 6th August 1945. From the dawn of consciousness until 6th August, 1945, man had to live with the prospect of his death as an individual; since the day when the first atomic bomb outshone the sun over Hiroshima, he has had to live with the prospect of his extinction as a species.[416]

6 August is also the day each year that Christians celebrate the Feast of the Transfiguration, when Jesus went up a mountain with his disciples Peter, James and John and became transfigured into pure light before them.

The Hiroshima bomb was transfigurative as well, shining 'brighter than a thousand suns', but its transfiguration was the opposite of the healing and redemptive light of Christ. Hiroshima began a permanent encounter with death immersion not only for those who saw the actual light of the atomic blast but for those who felt its psychic reverberations of guilt, death immersion and paranoia around the planet.

This death immersion jolted people from ordinary living into an existence of death-dominated life. Two cities of several hundred thousand people each were instantly pulverized without warning in a display of power that compelled the Japanese surrender and shocked the rest of the world.

For those who survived the bombing came the long-term encounter with the 'invisible contamination' of 'A-bomb disease'. First to die were those who succumbed to the effects of radiation sickness; then came the increases in leukemia deaths some four to five years later; then came the increases in the cancer rates and a variety of other physical complaints and illness; finally came the most traumatic blow of all – the discovery that *hibakusha* offspring – those having nothing to do with the war or the bomb – were experiencing abnormalities and genetic defects higher than the normal rate. Survivors thus have the sense of being involved in an ordeal with death which *began* on 6 August 1945, and has yet to end. As I pointed out, in the aftermath of the Harrisburg incident, the local townspeople there are beginning to come to the same awareness, that the accident at the Three Mile Island nuclear reactor was only the beginning of a trauma, the effects of which will only become apparent in years to come.

This physical death immersion causes what Lifton terms a 'psychic mutation' in the person involved. It is a transfiguration of his or her consciousness into an impaired state continually plagued by a dialectical tension with an awareness of death. Unable to integrate this awareness of mortality and death into the pursuit of life, most psychically close themselves off. This psychic numbing, however, does not erradicate the death immersion; ignoring the problem does not make it go away. Rather, psychic numbing results in guilt, feelings of counterfeit nurturance, paranoia, and finally an identification with the death force itself. This is why the Japanese use the term 'hibakusha' to describe the atomic bomb survivors. It means 'explosion-affected persons', ones who carry with them 'the identity of the dead'.

These changes are part of the survivors' ethos, of their way of being. As I observed, this death immersion radiated outward from the event, bringing the whole world into the awareness that, as Koestler points out, the entire human race, for the first time in history, now has the power to exterminate itself. Yet like most *hibakusha*, the world reacted with psychic numbing, the Americans getting caught up in the last stages of paranoia most of all – identifying with the bomb itself. It was seen as a weapon for all seasons, as the key to being able to dictate terms with the Russians, as a secret that they must monopolize and control at all costs. How much this psychological identification with the bomb was responsible for the ensuing cold war which developed between the US and the Soviet Union is difficult to say, yet the paranoia that accompanies such identification resulted in such intense mistrust, fear and alienation that the planet has been caught up in a spiralling nuclear arms race ever since, not only between the superpowers but between numerous smaller nations. Today, every city anywhere around the globe is threatened with the prospect of being engulfed in nuclear holocaust and being transfigured into another Hiroshima, for the superpowers, not content with being able to annihilate each other, have built up arsenals capable of killing each man, woman and child on the planet scores of times over.

Overkill in nuclear weapons is matched by similar overkill possibilities in nuclear reactors, for not only does each large reactor contain approximately 1,000 times as much radiation as the Hiroshima bomb, but the long term effects of *low-level* ionizing radiation are equally as disastrous as the *high-level* radiation release of a bomb. Many experts argue that long-term low-level radiation is *more* dangerous. In any case, a real connection exists between nuclear weapons and nuclear reactors – not only because the plutonium waste from reactors can be used to make nuclear bombs but because both weapons and reactors emit deadly radiation which kills and mutates human life and devastates the environment. Nuclear weapons and nuclear reactors, therefore, should not be seen as separate but as two sides of the same coin.

However, most people feel repugnance at the thought of nuclear weapons but not necessarily against nuclear reactors. The challenge for those who see their connection should not be that of psychological displacement, redirecting one's emotions and focus from the one to the other; rather, because both weapons and reactors involve the same substance – radioactivity – the more integrating psychological attitude should be that of extension: seeing them as organically bound together and therefore as equally dangerous.

This extension requires a psychic opening-up, an expansion of perspective embracing and integrating more truth. Such an opening-up not only allows one to appreciate the full dimensions of the problem, it can also spontaneously give impetus towards a drive to become part of the *solution* to the problem. This was a phenomenon observed among both *hibakusha* and concentration camp survivors. Amidst the many who were reacting to the death immersion by psychic numbing, a few were able to remain open, absorbing the death force through non-resistance to it, and thereby being capable of reconstituting themselves on the other side of the experience with a more fully integrated personality and sense of self-worth. Psychic opening-up allows one to see the totality of the experience at hand, the good as well as the bad, and, having seen the full extent of the evil, to embrace the good and identify with it.

Indeed, it is this psychological path through nothingness towards a reformulation of the Self on a higher and deeper dimension, that allows one to develop a sense of mission, 'survivor mission', as Lifton calls it. It is this mission that can transfigure the death immersion into a vehicle of light. It can allow one to become a source of help to people, a source of strength to those still caught up in the quagmire of guilt and paranoia because they have not yet opened themselves up.

This sense of mission is the foundation of the movement to stop nuclear power. Whatever else it might be, it is a movement which seeks to address the overall threat around us that is not only threatening the planet and the entire human species of the present generation but is producing radioactivity that will be equally threatening to life for generations living 10,000 years from now. It is a mission which seeks to address the most fundamental and primal concern we have, that of a threat to our very survival by the 'invisible contamination' of something we cannot see, taste, smell or feel – only fall victim to. The anti-nuclear movement is an expression of what Walter Canon, a neurophysiologist, has termed 'the wisdom of the body', a wisdom which utilizes anxiety as one of its most primary signals of danger to the organism.

One often hears the phrase from the atomic industry that it is time to 'set aside emotions' and appreciate the cost systems analysis that supports a nuclear programme. But whose emotions should be set aside? Why are we being continually asked to cut ourselves off from our own feelings in order to accept 'rationally' something we inwardly fear and intuitively do not want? Perhaps we should rather say: 'It is time to become more in touch with ourselves rather than less. It is time to rely more on our *body* sense because that is what radiation affects – not upon computer readouts which may support nuclear power before the accident, but once it has begun and the temperatures go off the scales can only print out question marks.' We must sort out our feelings, not eliminate subjectivity. We must be willing to be *moved* if that is what the facts demand and what our intuitive responses deem to be right.

The extension of nuclear weapons to nuclear reactors holds the key to putting us back in touch not only with our own selves but with the enormity of the imbalance that currently exists in the world. Nuclear weapons are hidden from public view and seldom seen except in military parades. They are easily distanced from, easily closed out of one's conscious existence. What actually happens is that their impact goes subterranean in peoples' minds, as a study by Michael Carey indicates.[417] He studied the death images and anxieties of the generation now in their thirties who have images and memories of atomic weapons, especially those derived from the civil defence drills held in schools throughout the US during the 1950s. The study has shown that the powerful images of human extinction implied by nuclear weapons produced such an inner terror that the child would usually repress it. This has produced a double-levelled psychic existence in almost all the people studied. On one level, there is the recognition that the world can be annihilated and that there is therefore no permanence or security; on another level, the conscious level, people, unable to cope with the fact of possible annihilation, attempt to pursue normal existence in spite of their underlying anxiety.

Nuclear reactors can serve to allow this anxiety to surface. There are very few people who live far from some aspect of the nuclear fuel cycle. In the US alone some 100 million people live within fifty miles of a nuclear reactor. While in weapons the threat is *potential*, in the rest of the nuclear fuel cycle the threat is *actual*, for at each stage radiation is released from normal operations, and the cancer or leukemia rates among both workers and public are increasing dramatically. While nuclear weapons can be dismissed with the rationale that they will never be used, therefore, nuclear reactors cannot be, for their impact begins from the moment they begin operation.

With rising cancer and leukemia rates, local populations are becoming uneasy, and active grassroots opposition is forming. This is particularly true for nuclear reactors and waste disposal sites. In Austria, even reactors which were already built were kept from beginning operation because of a plebiscite in which the Austrians expressed their opposition. In Gorleben, West Germany, the construction of a waste disposal site has been suspended because of 'political opposition', according to the governor who made the decision. Indeed, in every Western country with reactor programmes there are active grassroots groups seeking to stop those programmes.

The presence of the nuclear fuel cycle amongst us, then, even while subtly involving the workers at the plants and the local populations in death immersion, can simultaneously serve to re-sensitize us to not only the dangers of the fuel cycle but to the dangers of nuclear weapons as well. Once we are opened to one form of holocaust we can be sensitized to other forms. This dynamic demonstrates the truth learned by the *hibakusha*: that if one can begin to integrate the darkness it can invert our perspective and become the key to our enlightenment. This is to say, that when the nuclear weapons programme was extended to nuclear reactors, the energy dynamic has been such that its effects through radiation exposure have touched enough people to begin a critical mass of opposition that is beginning to roll back the entire nuclear complex, particularly in the US in the aftermath of Harrisburg and the Karen Silkwood case. People are being touched by radioactivity; they are being opened to its effects; they are moving to stop it. It is almost exactly as Einstein said, that the only thing that would stop the abuse of nuclear power, specifically in his mind the nuclear weapons race, would be the taking of the truth to the 'village square'. Once this was done, he said, and the common people knew the truth about the situation, there would begin a *groundswell* of opposition so intense that the governments would be forced to respond.

V

An Alternative Vision

Fundamental to any groundswell of opposition to nuclear power being successful is that it be balanced with a positive alternative. It is not enough to take a stand *against*; one must be equally certain what one is *for*. This means that while taking to the 'village squares' the message of the destructiveness of the nuclear weapons/reactor complex, there must be an equally forceful demonstration of alternative ways of generating the energy we need as well as alternative ways of relating to each other as human beings. Without this the movement will become counterproductive, remaining only negative and leaving a vacuum in its wake.

A positive alternative to nuclear power must be twofold: it must demonstrate viable alternative energetics, thus offering a clear substitute to nuclear reactors, and it must personify a mode of interpersonal relationship based on nonviolence, thus creating an atmosphere of trust in which nuclear weapons could be laid down. A detailed description of either one of these is beyond the scope of this book. They are being offered as *directions* in which we must explore, for by their very nature they are not amenable to hard and fast delineation.

A. THE SOFT ENERGY PATH

The traditional rationale for nuclear power is that it is a necessary component for an economic system which seeks, on the one hand, to maintain an undifferentiated growth-rate and corresponding energy demands, and on the other hand, which is encountering increasing economic, political and evironmental scarcities of oil and gas. The energy problem is defined as how to increase domestic energy supplies to meet extrapolated homogenous demands, particularly in the light of the fact that a traditional reliance upon fossil fuels is becoming an increasingly unviable option.

When it is suggested that a combination of conservation of existing energy sources with a utilization of renewable sources of energy could fill the gap much more effectively, the official reply invariably is similar to that given by Sir Brian Flowers in an article entitled 'Nuclear Power: A Perspective of the Risks, Benefits and Options':

> Alternative sources will take a long time to develop on any substantial scale. . . . Energy conservation requires massive investment in new industrial processes and in urban technology, and can at best reduce somewhat the estimated growth rate. Nuclear power is the only energy source we can rely upon at present with any certainty for massive contributions to our energy needs up to the end of the century, and if necessary, beyond.[418]

Amory Lovins has demonstrated, however, that to satisfy the undifferentiated growth rate in both the economy and the energy supplies projected by the governments and accepted by Sir Brian Flowers, nuclear power is simply incapable of fulfilling the task: 'it is not economically capable of proving significant substitution for oil even if given infinite time.'[419] To substantiate this assertion he points to the following observations:

> To meet with nuclear power a quarter of the officially projected demand for energy in the US by the year 2000, it would be necessary to order one new 1000 megawatt reactor every five days from now till the end of the century. This assumes that each new unit of nuclear electricity will displace two units of primary fossil fuel throughout the economy.
>
> To meet one half of the official Japanese energy requirements at the year 2000 with nuclear power would require ordering one new 1000 megawatt reactor every twenty days for the next twenty-one years. Even if this was done, fossil fuel imports would still increase by 73%. This estimate also assumes that each unit of nuclear electricity will displace two units of delivered fossil fuel.
>
> In non-Communist Europe, if the official estimates hold of a 3% per annum growth rate until the year 2000 and from thence a 1.6% growth rate until 2025, and if the estimates hold as well of there being 1,000,000 megawatts of nuclear

power by the year 2000 and 2,100,000 megawatts by 2025, then non-nuclear energy requirements would still increase 89% between 1975 and 2000 and 150% between 1975 and 2025.

In order to replace half of the world's 1970 consumption of oil and gas with nuclear power would require over 1,600,000 megawatts of nuclear electricity – more than the 1970 combined total generating capacity of the US the USSR, the EEC and Japan.

Lovins concludes:

It may be said that without nuclear power, all these examples would look even worse. But their point is that even prohibitively large nuclear programmes cannot possibly keep up with recent official projections of energy demand. . . . Rather than pretending nuclear power be the answer, we should accept the need for the kinds of dramatic efficiency improvements . . . without which no supply policy can make sense. We should then ask, freed from uncritical acceptance of the ability of nuclear power to solve the oil problem, whether other investments might not buy more or quicker relief from oil dependence – lest, having like Lord Flowers dismissed alternatives as slow, conservation as costly, and both as inadequate, we choose a nuclear future that is simultaneously slow, costly *and* inadequate.[420]

The positive alternative to nuclear power is therefore twofold: first, it emphasizes conservation and efficiency improvements; and second, the development of alternative energetics.

It is estimated that approximately 50% of all the energy generated in most Western European countries is wasted. In the US the percentage is closer to 65%. This is due to inefficient machinery, inefficient use of the energy generated, improper insulation of homes, thus allowing most of the heat to escape, and simple waste. Because of the enormity of the waste, simple conservation of the energy already being generated could not merely increase the average efficiency of primary energy use by a few percentage points but by *severalfold*. This can be seen from the following examples offered by Lovins, all of which are based on conservative economic and technical assumptions:[421]

A study done by Gerald Leach and the International Institute for Environment and Development has indicated that the

United Kingdom could *treble* its Gross National Product over the next fifty years with no increase in its primary energy requirements, utilizing only technical efficiency improvements that are within the present art and which are all essentially cost-effective at present British energy prices.[422] Another study done by the Earth Resources Research group has demonstrated that if cost-effectiveness is judged against long-run marginal costs, then technical efficiency improvements can achieve a further doubling of primary energy efficiency. This would mean that primary energy use in 2025 would be 0.42 times its 1975 level, industrial production would be 2.3 times its 1975 level, and real Gross National Product would be 2.9 times its 1975 level.

In Denmark, similar studies have indicated that primary energy use could be reduced 52% between now and 2025, chiefly by technical improvements, while living standards would continue to improve.

In France, conservation and technical efficiency improvements could yield a situation where the Gross National Product would increase between now and 2025 at the officially projected rate of between 2% and 4½% while yet by 2025 only having used 10% to 23% *less* end-use energy than in 1975. There could be a substantial rise in the standard of living, therefore, while using less energy than that used in 1975. Another study has indicated that while the official French government projection is for an energy growth 105% more than current levels, almost exclusively technical improvements could hold the 1975-2000 end-use energy growth to a total of not more than 23% above current levels.

In West Germany, where official forecasts project an increase in end-use energy by 2.42 times by the year 2025 and in primary energy growth by 3.41 times, an outline study recently completed showed that with only straightforward technical improvements there could be the officially projected economic growth rate until 2025 without any increase at all in the end-use energy growth rate.

An official Swedish study has found that by 2015, all Swedes could enjoy the same standard of living as is now enjoyed by the top 10% of the population with an end-use energy growth of only 25% and a primary energy growth of 38%.

In the Netherlands a study has shown that conservation

and technical efficiency improvements could have cut the 1975 level of energy expenditure without in any way cutting industrial activity. Studies in Switzerland indicated similar conclusions.

In the United States, a National Academy of Sciences panel has asserted that the US could double its Gross National Product by 2010 with only an 8% total increase in primary energy use and an actual 11% decrease in primary energy use if modest life-style changes were implemented. In 1978, a Department of Energy study found that with no changes in values or lifestyles, California alone could treble its economy by 2025 with only a 20% increase in end-use energy, if modest technical improvements were made. If conservation and technical improvements were systematically employed, then there would be the same economic growth but with a 22% decrease in end-use energy levels.

In Canada, a study for the Science Council of Canada found that if Canada was to increase its population by 75%, more than double its industrial production rate, and experience similar advances in the other sectors of the economy between now and 2025, it could still reduce its end-use energy needs by 14% of the 1975 level if technical efficiency improvements were made.

What emerges from the above studies greatly substantiates the assertion made by Roger Sant, Assistant Administrator of the Federal Energy Administration of the US that 'By 2010, conservation could save an amount (of energy) equal to our present useage, making it the most important step we can take. For the most part, energy-efficiency investments in all sectors are the least expensive new sources of energy we will ever find.'[423] John Berger's observation is equally forceful: 'The energy bonanza is conservation. Its potential is so enormous that – if tapped extensively – it makes the building of uranium fission plants completely unnecessary.'[424]

The challenge of conservation and technical improvements is that of *re-defining* the energy crisis virtually every country is in. While up till now the assumption has been that of how to get more energy to meet projected homogenous demands, 'we should re-define the problem,' argues Lovins, 'as *how to meet heteronomous end-use needs* – the many different tasks we seek to do with the energy *– with a minimum of energy supplied in the most* *effective way for each task'*.[425] It is a redefinition, then, which seeks a *minimum* amount of energy rather than a maximum amount, which seeks the right tool for the right job within the context of economic growth and a continual rise in living standards.

It is within this minimalist approach to energy which seeks the right tool for the particular job that the soft technologies can be employed. This soft energy path is defined by Lovins as 'diverse renewable sources that are relatively understandable to the user and that are matched in scale and energy quality to their application'.[426]

The soft path seeks those energy sources that nature provides naturally, those energy sources that will always be with us because they are always being renewed. It is the opposite of the hard path currently employed that sees the earth as a resource to be exploited, with minerals to be used up such as oil and uranium, and then when they are depleted to begin the search for other sources.

The cornerstone of the soft energy path is the sun, a gigantic source of light which bathes the earth each day in an enormous energy field. The amount of energy we receive from it is so enormous that it is far beyond any currently conceivable energy needs. While the oil companies and the nuclear industry and their spokespersons in the government speak of an energy crisis, therefore, the surface areas of the US alone receives approximately *nine thousand trillion* kilowatt hours per year in solar energy, an amount that is roughly 600 times the current American energy use.[427] *There is only an energy crisis because the energy companies and the governments are looking in the wrong places.* They are seeking energy from sources that can be used up and depleted rather than using sources that can never be used up; they are selling energy from sources that are finite and therefore expensive – thus maximizing their profit, rather than harnessing energy that is infinite in its renewability.

It is estimated that 0.7 kilowatts of energy from the sun falls on every square metre of land around the earth during daylight hours. This is equal to 64 watts per square foot. Indeed, so much sunlight falls that even in areas such as the rainy parts of northwestern US and in much of Britain the equivalent of three times the average household electricity use can be collected from the rooftops by solar collectors.[428] In places such as the deserts of Saudi Arabia, enough solar energy reaches the earth to equal the world's entire reserves of oil, gas and coal.[429]

The technology for using the sun for both heat and electricity is available, is safe and can be used with a minimum amount of damage to the environment or the public. Solar plants will, of course, add to the heat burden of the earth by increasing the absorption of solar energy and then converting it to electricity with more losses of heat. This is a problem it shares with all power technologies, however, and is insignificant in comparison to the radioactivity releases of nuclear reactors.

Unlike energy generated from fossil fuels or from uranium, energy from the sun is virtually inexhaustible. Furthermore, because solar plants burn no fuel, they will not pollute the human and natural environment with either particulate or radioactive discharges. Solar power thus has an elegant simplicity and cleanness which no other energy technology can match. Another asset of its use is that most of the solar power available comes in the afternoon when more energy is demanded of the power grids than at any other time. Most importantly, solar energy is everywhere so that each home, school, hospital, business, factory and industrial plant can build the collectors and generators specifically suited to its need. Solar energy does not require a huge energy conglomerate to generate the electricity or heat the public needs and then sell it to the public. *Solar energy makes the selling of energy for profit anachronistic*.

Because the sun is diffuse, uncontrolled by the power companies, and therefore variable in its daily intensities, particularly given seasonal and geographical differences, it is by nature an energy source peculiarly suited to decentralized case-by-case deployment. This diversity adds somewhat to the cost of solar energy but not prohibitively so. It certainly has not inhibited the technical advances in the field. Solar technology for home heating and cooling has been available since the 1950s. If this technology were widely utilized today, the citizens could enjoy energy independence from the utility companies which are currently selling them everything they need.

The availability of solar technology as well as its viability is increasingly being recognized by even agencies dedicated to nuclear energy. In 1973, for example, the Atomic Energy Commission published a study concluding that 'Ultimately, practical solar energy systems could easily contribute 15 to 30 per cent of the nation's energy requirements.'[430] The study also found that solar energy could provide 30% of America's heating and cooling needs by the year 2000 and that solar photovoltaic cells could be competitively priced by the mid 1980s. As one might expect, coming from the AEC, these estimates were only grudgingly made and extremely conservative.

What is important to note is that solar energy means far more than just the sun. The daily heat from the sun not only in turn heats the planet but provides the means for all plant growth, for all weather conditions, and for the movement of the currents in the oceans. It is thus the source of all wind energy, wave energy, hydroelectric power, geothermal power, biomass power. It is even responsible for fossil fuel energy.

There is so much energy available in the renewable energy sources that studies for over fifteen industrial countries have not yet found one where indigenous renewable sources of energy cannot meet demands and even provide a long-term energy surplus. The following countries can serve as examples:[431]

In the United Kingdom, renewable energy sources could provide 18% of energy needs by the year 2000, 68% by 2025, and 100% by 2050.
In Denmark: 44% by 2000, 83% by 2025, 100% by 2025 if desired.
In France: 14% by 1985, 28% to 39% by 2000, 67% to 100% by 2025.
In Sweden: 59% to 81% by 2000, 100% by 2015.
In Canada: 100% by 2025.
In California: 86% to 100% by 2025.

In most cases, the transition to renewable energy sources can be facilitated by regional co-ordination such as between the Scandanavian countries, between neighbouring Canadian provinces, and between California and the northwest US.

Lovins is quite categorical in stating that, 'The environmental and sociopolitical advantages and the potential resilience of improved efficiency, soft technologies, and (to a lesser extent) transitional fossil-fuel technologies are not open to serious dispute. Nor is technical availability. . . .'[432] The only major point at issue is that of deployment rates and cost as compared with hard path currently being employed of central-electric systems such as nuclear power.

It is important to bear in mind, says Lovins, that, 'No energy programme can eliminate the use of fossil fuels over a transition of some decades; the question is which can minimize that use subject to investment and other constraints.'[433]

It is perhaps within the context of attempting to relieve dependence upon oil in the coming years that the most astounding deficiency of nuclear power shows up. Nuclear power is used exclusively to generate electricity, and yet our energy supply problem is more than 90% a problem *not of electricity* but of heat and portable liquid fuels. Therefore, for the energy companies and the governments to look to nuclear power as the solution to the oil crisis is to be looking in completely the wrong direction again.

Just how wrong a direction the nuclear option is has been pointed out by Vince Taylor in his book, *Energy: The Easy Path*.[434] As a thought experiment, Taylor assumes that every single oil-fired plant – whether gas turbines, diesels or steam – are replaced overnight in the US, Western Europe and Japan by 146,000 megawatts of nuclear plants operating at 65% capacity. Of this amount, 54,000 megawatts are in the US, 45,000 in Japan, 9,000 in the UK, 7,000 in France, and 3,000 in West Germany. Even with this enormous substitution, however, oil consumption is reduced by only 12% overall, although in the most critical area of oil – importation, consumption is only reduced from 65% to 60% Japan experiences the most dramatic savings in oil, reducing imports by up to 29%, although 90% of the oil that must still be consumed must be imported. Taylor's nuclear substitution would reduce the consumption of imported oil in Britain from 94.9% to 94.1% with an overall oil consumption reduction of 14% in France from 95.6% to 95.1% with an overall reduction of 10% and in Germany from 92.5% to 92.2% with an overall oil consumption reduction of 4%. These changes are as small as they are because Britain and West Germany depend primarily on coal, not on oil, and France upon hydroelectric power. The reason for the complete inadequacy of nuclear power to solve the oil crisis is that so little oil is used in power stations. It is used primarily to generate heat and for liquid fuel. This is not an area nuclear power can be any help in as its limitation is to a power plant that produces electricity. 'The most ineluctable limit to the potential role of nuclear power, then,' says Lovins, 'is set not by fiscal, political, or logistical constraints (though these are certainly formidable), nor by uranium scarcity, but by the lastingly small market that it is able, even in principle, to penetrate: the market for baseload electricity for electricity-specific end use.'

Against this poor substitutability of nuclear power should be seen the renewable sources. Because they are of small scale, are not dangerous, and are easily understandable, they have far shorter technical lead times than nuclear plants, which take upwards of ten years to build. A one megawatt wind machine, for example, can be delivered within 8 to 12 weeks.

Moreover, because the devices are small and simple for the user, they can be sold to literally millions of households rather than to a few utility companies. Renewable technologies are therefore suitable to the broader market of normal consumer goods rather than to the slower process of a technology geared towards specialized narrow markets.

Finally, renewable energies are suitable to technical diversity, thus embracing dozens of kinds of technologies whose institutional rate restraints are largely independent of one another.

Lovins concludes:

> Such considerations suggest that relatively dispersed investments in quick, small units should be able to relieve oil dependence at least as quickly as centralized, long-lead-time investments in large units, and arguably a good deal faster. The cost advantage of the former over the latter is demonstrably large and robust, and the cash flow advantage larger still: an investment that takes ten years to build and thirty to pay back requires many times as much working capital, even if the specific investment is the same, as one that takes a month to build and a few years to pay back. Thus in a real market, with gross imperfections (institutional barriers,) and asymmetric subsidies removed, efficiency improvements and soft technologies would easily win over marginal central-electric systems on grounds of lower private internal cost.[435]

B. THE WAY OF NONVIOLENCE

Even as the soft energy path of conservation complemented with the utilization of renewable energy sources is to be seen as a clear and viable alternative to the hard path of centralized power stations and nuclear reactors, so the new emotional and mental attitudes that would complement a more organic economic connection with the earth must be striven for against the backdrop of the current dehumanizing technology and mindless bureaucratic imprisonment most people feel enmeshed in. The hard energy path is matched by a hard world-view of energy scarcity, of competition for these scarce and dwindling sources, and of a resulting alienation not only from our competitors but from the earth itself. The soft path, on the other hand, because it uses those sources which are always renewed, results in an attitude not of competition and alienation but of co-operation and individual empowerment. It is a path that can put us back in touch with the sense of humanness and self-worth that the hard path has cut us off from.

The soft path, then, appeals to a very deep sense within us that seeks to unite with the earth and each other. The contradiction that exists is that it is being articulated in the midst of an age of overkill possibilities and in the midst of nations which seem intent on the hard energy option, no matter what the risk or the economic absurdity of doing so. Any residual humanness within us, therefore, is forced to interlock with what US National Security Advisor Brzezinski calls a 'technotronic age' reality of instantaneous communications systems around the globe via satellite and telecommunication links which are interconnected with an instant access to information via computer technology. The technotronic society determines human worth by the measure of silicon chips. To interlock with this hard path dehumanizes the one interlocking to the degree s/he is dependent upon its machines for self-identity. This causes what L. L. Whyte calls a 'fundamental division between deliberate activity organized by static concepts, and the instinctive and spontaneous life'.[436]

This 'fundamental division' has reached such an extent that the 'static concepts' of the technotronic age have taken on a life of their own and are currently dragging modern humanity along behind them. This is causing 'behaviour patterns unrelated to organic needs' and an 'uncontrolled industrialism and excess of analytic thought' that has been completely detached from the 'catharsis of rhythmic relaxation or satisfying achievement' that each human being needs to maintain inner balance and a sense of rootedness and self-value.[437]

Put more simply, we have been turned into consumers of mass-production systems, conditioned to want things we do not need, imbued with a relentless passion for quantity rather than quality that forces us to buy products designed with built-in obsolescence, and bombarded with commercials and advertisements which inform us of the next fad or product necessary to maintain the facade of modernity. Observing this phenomenon from India, Ananda Coomaraswamy wrote: 'The West is determined, i.e., at once resolved and economically "determined", to keep on going it knows not where, and calls the rudderless voyage "Progress".'[438]

The East, too, however, has been caught up in this 'rudderless voyage'. One has only to look at post-war Japan and at the current open-door policy in China in the Orient and at the newly emerging Third World nations throughout Africa and Asia to see that all appear intent on joining in. It is a race to modernize at any price, to be armed with the most sophisticated of conventional and, if possible, nuclear weapons, to be equipped with the most modern of industrial complexes fed through the energy grids of centralized nuclear power plants. The hard-energy path of the Western economic system, now spread all over the world, has produced as much psychological and emotional imbalance as economic dislocation and political competition and alienation. Indeed, Coomaraswamy is right in observing that, 'We are at war with ourselves and therefore at war with one another. . . . Man is unbalanced, and the question, Can he recover himself? is a real one.'[439]

Thomas Merton, a Western religious thinker, puts the matter perhaps most clearly in stating that what we are confronted with

> is a crisis of *sanity* first of all. The problems of the nations are the problems of mentally deranged people, but magnified a thousand times because they have the full, straight-faced approbation of a schizoid society, schizoid national structures, schizoid military and business complexes . . . and schizoid religious sects.[440]

This 'crisis of sanity' is not, however, merely the result of a technotronic society. It is as much the result of an explosion of knowledge that has outpaced the individual's ability to keep pace. Alvin Toffler calls this 'future shock', a psychological dislocation produced when the knowledge and technology that is outpacing our ability to integrate it explodes our present from the vantage point of the future. And yet, he says, 'the shattering stress and disorientation that we induce in individuals by subjecting them to too much change in too short of time' is not a future possibility but a present physio-biological condition which he calls 'the disease of change'.[441]

Toffler offers a further insight which is important, that 'the *rate* of change has implications quite apart from, and sometimes more important than, the *direction* of change'.[442] This theory is derived from William Osborn's observations concerning cultural lag, that social stress arises out of an uneven rate of change in different sectors of the social structure.

At base, therefore, what future shock consists of is a *time phenomenon* which Toffler exemplifies in the following way:

> It has been observed that if the last 50,000 years of man's existence were to be divided into lifetimes of approximately 62 years each, there have been about 800 such lifetimes. Of these 800, fully 650 were spent in caves. Only during the last 70 lifetimes has it been possible to communicate effectively from one lifetime to another – as writing made it possible to do. Only during the last six lifetimes did masses of men ever see a printed word. Only during the last four has it been possible to measure time with any precision. Only in the last two has anyone anywhere used an electric motor. And the overwhelming majority of all the material goods we use in daily life today have been developed within the present, the 800th, lifetime.[443]

Commenting on this, Hans Küng states that:

> The upheaval of our lifetime . . . must be regarded as the second great break in the history of mankind, the first being the invention of agriculture in neolithic times, and the transition from barbarism to civilization. Now, in our times, agriculture, which constitutes the basis of civilization for thousands of years, has lost its dominance in one country after another. And at the same time the industrial age, begun two centuries ago, is passing. As a result of automation in the progressive countries, manual workers are rapidly becoming a minority and a superindustrial culture appears on the horizon, perceptible only in outline.[444]

What this means to someone like Kenneth Boulding, an economist and social thinker, is that 'the world today is as different from the world in which I was born as that world was from Julius Caesar's. I was born in the middle of human history, to date, roughly. Almost as much has happened since I was born as happened before.'[445]

The *rate* of change causes its own stresses and dislocations. When this is coupled with the *direction* of the change going into

Overkill and the plutonium economy one can begin to understand why it is that the sense of not being at ease with technical civilization has become universal despite the fact that we seem intent on embracing it.

What we are experiencing, then, is the culmination of classical world politics within the context of the climax of an industrial development begun two centuries ago in Britain. Nuclear weapons and nuclear reactors merely dramatize the transitions that have been occurring and the alienation that has been wrought from the very life the industrialization was supposed to serve and the classical politics to protect. The bombing of Hiroshima set in motion psychological reverberations of guilt and paranoia that have only deepened the alienation of the future shock the world was already in, transforming the world into a society of nations on a permanent war footing. It has become a permanent war economy, however, that is capable not only of making war but of exterminating the human species.

The symbol of future shock, therefore, the impetus for not only the dire consequences of *attempting* change without reflection but of *directing* change without proper balance is the nuclear weapons/reactor complex. It is a phenomenon and a symbol which cuts history in two like a knife with a transformative power over human existence that has given us the power of ultimate destruction.

The paradox within this fact is that it is within the depths of darkness that one must search for light. It is at the point where future shock bears down upon human beings the hardest in the form of a technotronic reality spewing forth plutonium waste that one must begin the long slow process of building the soft energy alternative. Lifton is quite penetrating on this point: 'The paradox,' he says, 'is an ultimate one. The existence of weapons that can annihilate man and his history could also, however indirectly, be a stimulus toward a deeper and more humane grasp of the same. Hiroshima was the prelude to all this – an expression of technological evil and madness which could, but will not necessarily, be a path to wisdom.'[446]

This 'path to wisdom' must be a path that leads away from the violence that nuclear weapons and reactors imply to their polar opposite – towards *non*violence.

Much can be said about nonviolence. I would define it here as the path each of us is challenged to take to resensitize ourselves with our emotions, our bodies, our connections with each other and the earth. Nonviolence is the reconnection each one of us feels between his or her *individual* life and the *source* of all life.

In discussing the soft energy path I observed that this ultimately means becoming symbiotically relinked with the energy sources mother nature provides in her own processes of harmony and motion. It is a path which seeks to use appropriately those energy sources which spring spontaneously from the movement of the sun and the motion of the earth, energy sources that will always be with us because they are always being renewed, whether we use them or not.

We will only be able to redirect the course of history towards the soft energy alternative if enough people are personally in tune with the rhythms of nature with which we are seeking to relink. We will only be able to apply energy to technology appropriately if we have regained our sense of equilibrium enough to know what 'appropriate' means.

Fundamental to the notion of the soft energy path is the spirit of attunement with oneself. This attunement is the basis of nonviolence: a realization of connection within oneself with the source of life itself. Nonviolence, therefore, is not a means to achieve unity; it is a manifestation of a sense of unity already there. We can only act in accordance with what we are. If we choose to manifest the darkness within us, violence and dehumanization result; if light, then goodness will result. Both forces are within us just as both the hard and soft paths are equally available outside us. Nonviolence sinks beneath this duality and relinks with that cornerstone within us that is absolute. From this place of connection comes awareness of what is and the ability to be compassionate. Without balance internally there will be no balance externally.

While nonviolence is a manifestation of an inner sense of connection, it is not achieved by an inward journey that produces isolation. Rather, the inward journey for attunement creates a deeper sense of understanding and attachment both to other people and to the earth, for they are equally parts of the same life-force that gives each seeker his or her life. Nonviolence, therefore, is as communal as it is individual; *one's own liberation, one's own quest for balance, is directly linked with that of others*.

Plato taught this, saying that 'to philosophize and concern with politics is one and the same thing, and to wrestle with the sophist means at the same time to defend the city against tyranny'.[447] This truth can be seen amply in the life and death of Socrates, Plato's teacher. His dictum 'Know Thyself' meant equally 'Help Others Know Themselves'. For this, perhaps the greatest philosopher in the West was charged with corrupting the youth and with spreading anarchy in the city – and was given poison to drink.

This connection between one's own liberation and that of others is the bedrock of nonviolence and has been taught by all great nonviolent revolutionaries. The person whom Mahatma Gandhi called an example of 'nonviolence par excellence', Jesus of Nazareth, lived out this truth above all things. For him, the greatest of all models of behaviour was to love your neighbour as you love yourself, meaning never to see yourself as isolated from people but as inseparably interconnected with them. In this sense, as Henry Wieman puts it, Jesus 'split the atom of human egoism' with a living testimony that 'the creative power lay in the *interaction* taking place between individuals, transforming their minds, their personalities, their appreciable world, and their community with one another and with all men'.[448]

In an age, therefore, in which the world is divided into warring camps of competing and alienated nation states, each armed to the point of being able easily to kill both the enemy and themselves; in an age in which multinational corporations and utility companies are willing to sacrifice atomic workers and nearby populations to the plutonium economy as 'radiation fodder' to keep the economy going, Jesus, Gandhi, Martin Luther King say 'Love your enemies. Humanity is One.'

This is such a fundamental concept to the whole notion of nonviolence, according to Gandhi, that 'there is no limit whatsoever to the measure of sacrifice that one may make in order to realize this *oneness with all life*'.[449] Gandhi said this, knowing that this sense of oneness was the 'antithesis' to modern civilization's dictum: 'Increase your wants. See nature and other people in terms of how they can benefit *You*.'

As the antithesis of modern nuclearized society, the first principle of active nonviolence is *non-co-operation* with this society's dehumanization, with its injustices, and perhaps most importantly, with the commitment it has to untruth. This non-co-operation negates the injustice the dehumanization, the untruth, by refusing to give them credibility. Moreover, non-co-operation refuses to accept the violence that is necessary in order to keep such a society stable, insisting that the forces of injustice and untruth which dehumanize people in society can only be enforced and perpetuated by violence.

Here again, however, nonviolence balances the inner refusal to co-operate with any form of violence or dehumanization with an equally important demand that the person who does not co-operate should not cut him or herself off from the people still caught up in the violence. Nonviolence turns non-co-operation with evil into the saviour of the people involved with the evil because it not only continues to see the connections between all persons involved but refuses to accept the *finality* of evil. In refusing to co-operate with evil, therefore, nonviolence steadfastly refuses to make the person committing the evil into an enemy. Nonviolence has no enemies, for it clings firmly to the belief that deeper than the evil a person may be caught up in is good, that beyond the circumstances that have cut him or her off from this good are those forces within him or her which, if tapped, will cause a conversion from identification with the evil to a realization of the power and unity of doing the good. As Gandhi puts it, 'If love or nonviolence be not the law of our being, the whole of my argument falls to pieces.'[450] For nonviolence, therefore, deeper than hatred is love, deeper than cruelty is kindness, deeper than oppression is liberation.

In her book, *The Human Condition*, Hannah Arendt explains this point further. Nonviolence, she says,

> recognizes that sin is an everyday occurrence which is in the very nature of action's constant establishment of new relationships within a web of relations, and it needs forgiving, dismissing, in order to make it possible for life to go on by constantly releasing men from what they have done unknowingly. Only through this constant mutual release from what they do can men remain free agents, only by their constant willingness to change their minds and start again can they be trusted with so great a power as that to begin something new.[451]

Merton emphasizes this most crucial point somewhat differently, pointing out that, 'in the use of force, one simplifies the situation by assuming that the evil to be overcome is clear-cut, definite, and irreversible. Hence there remains but one thing: to eliminate it.' Nonviolence, however, realizes that 'to punish and destroy the oppressor is merely to initiate a new cycle of violence and oppression. The only real liberation is that which *liberates both the oppressor and the oppressed* at the same time from the same tyrannical automatism of the violent process which contains in itself the curse of irreversibility.'[452]

True nonviolence, then, born out of a sense of one's own inner unity and connection with the source of life, strives to see this unity and connectedness in the other people whether or not they can see it themselves. It is a posture of love that refuses to make enemies, choosing, rather, simply not to co-operate with any manifestation of evil.

To attain this sense of inner balance and centredness and the courage not to co-operate with the forces of disharmony, whether in yourself or the other person, has become the most crucial challenge of our time. We have developed a weapons system that can annihilate humanity scores of times over; we have produced a nuclear fuel cycle that in its normal operations is killing people by the thousands with leukemia and cancer and even goes so far as genetically to damage the unborn. Against this violence one cannot return violence, for as Gandhi points out, 'You cannot successfully fight them with their own weapons. After all, you cannot go beyond the atom bomb.'[453]

This is an important point for those in the movement against nuclear power who believe in violence to reflect upon. Against a society armed and willing to use nuclear weapons, no conventional arms would be successful, for they are inferior in kill power. Yet nuclear weapons against nuclear weapons would almost surely mean double suicide, leaving both revolutionary and reactionary, terrorist and society equally destroyed. Before Hiroshima, violence against violence was a credible option perhaps; now it has become irrelevant and counterproductive. Gandhi is right in saying that 'In this age of the atom bomb unadulterated non-violence is the only force that can confound all the tricks of violence put together.'[454]

The only recourse to induce disarmament, therefore, is not to *fight* society but to *transform* society, and this can only be done through the nonviolent posture of not co-operating with the violence, the injustice, the untruth, while simultaneously seeking to liberate those persons perpetuating it. This may seem to be matching David against Goliath, but if one realizes with Gandhi that 'nonviolence is the only thing the atom bomb cannot destroy . . .'[455] then one becomes empowered with a force that will in time be able to diffuse the hate and paranoia that gives reason for the atom bomb. Neither atoms for war nor atoms for peace would exist without the accompanying attitudes of paranoia, hate, fear, and alienation. Nonviolence directs all its energy at eroding this negative support system in people's hearts and minds, believing that if the feelings of paranoia are replaced with a willingness to co-operate, if the hate is replaced with a feeling of fraternity, and if the alienation is replaced with a sense of connectedness with the earth and with other people, then the weapons will be laid down and the reactors will be entombed.

Recall the discussion at the conclusion of the section on nuclear weapons where the co-operation-for-conflict theory of Bigelow was offered as a possible explanation for the phenomenon of the age of overkill. It was suggested that deep in our evolutionary heritage is a striving for dominance, an adherence to territoriality, and a competitive fear of anyone not in 'your' group. Groups, therefore, began to co-operate within themselves in order to be able better to compete with other groups for limited resources and food supplies. Gradually, however, the advantages of inter-group co-operation were seen, a difficult step to take, to be sure, as it required making alliances with former enemies, but one which allowed each individual group in the alliance to feel stronger and more secure. The result was that the contending armies became larger, the weapons more devastating, the casualties higher. But, and this is of critical importance for our purposes here, the areas within which relative peace prevailed also increased. Freud was also quite struck by this paradox, that 'it is war that brings vast empires into being, within whose frontiers all warfare is proscribed by a strong central power'.[456] This is to say, that the larger the unit, the greater the internal peace. Pax Romana is an example of this, where a strong central authority could maintain an internal peace within the perimeters of its political boundaries. While there are problems with this, first, that such stable units never last for long and, second, that the larger the unit the more destructive and violent its downfall, Freud as well as Bigelow assert that this observation, that the larger the unit the greater the area of internal peace, is the only salvation from continued warfare among the nations of the world. *To end warfare necessitates making the perimeters of the peace as large as the dimensions of any possible conflict*. What this would necessitate in our day, given the fact that the entire planet is being threatened, is some type of planetary government within which the peoples and the nations of the world would feel secure enough to lay down their right to bear arms against one another. Such security would have to be predicted upon at least two things: first, that each individual and group would be guaranteed certain basic human rights that could not be violated. The establishment of an international court of judicature to clarify and protect these rights, accessible to all, would facilitate this process. The second security is that against the ravages of hunger and of being deprived of the resources necessary for human well-being. War has always resulted from these two sources. People have gone to war and made revolution when they believed their human rights and/or existence was being unduly threatened; and people have gone to war over resources and food supplies. Whatever planetary co-operation established, therefore, to end the threat of global annihilation, these two areas of human need must be addressed.

For millennia, human beings have fought over these things, each time bringing into battle more sophisticated tactics and more potent weapons. In the end, it has been the might of the one side over the other that has established the 'right'. In the light of nuclear weapons arming all sides, however, Bigelow implores us to understand that 'The time has come where our survival *as a species* depends on whether we have brain enough to understand the words: "United we stand, divided we fall".'[457]

Given the fact that our co-operation-for-conflict drive is so deeply engrained in us as to be virtually an instinct, the chances that we can overcome our duality and embrace each other as equals, transcending the distinctions between Marxist and capitalist, between black and white, between haves and have nots, between one religion and another – all those things for which we have in the past been willing to arm ourselves – is slim. As the anthropologist Ardrey points out, our species was born with a weapon in our hands. Our predicament, therefore, is similar to that of the Pliocene gorilla. The tree bough was the focus of its experience even as the weapon has been a central focus of ours. It provided the gorilla with both fruit for nourishment and the means of locomotion, dominating its existence to the point of causing the specialization of gorilla anatomy: the hook-like thumbs, the powerful chest, the long arms, the weak and shortened legs. 'The bough,' says Ardrey, 'was the focus of gorilla tradition, gorilla instinct, gorilla security, gorilla psyche, and of the only way of life the gorilla knew.'[458]

If humanity is voluntarily to give up the right to bear arms against one another, depriving itself of the contest of weaponry, our equivalent of the gorilla's bough, then the challenge before us, says Ardrey, is nothing short of 'a new mode of existence'. Whether this will occur, given the limited time in which it must happen before irreparable damage is done to the planet as well as to human beings, is highly problematic, Ardrey positing that 'given access to our traditional materials, we shall proceed with alacrity to blow up the place'.[459] Einstein, for his part, tended to agree with Ardrey's judgment, saying that because the species had made a quantum leap in technology when it split the atom, changing everything in the world – except for human consciousness, we were heading for what he termed 'unparalleled catastrophe'. Both men offered the hope, however, that a remnant, mutated to be sure, will probably survive the holocaust and be able to begin anew.

It is against this Goliath, it is against this deeply engrained aggressiveness and competitive mistrust that seems to compel the use of weapons and violence, that nonviolence comes. To a world divided, it speaks of unity; to people caught up in oppression and tyranny it reacts with liberation and love; and to persons alienated from their own beings it manifests connectedness with the source of all life. It does not seek to kill Goliath, but merely through non-co-operation with his aggression to convert his soul.

The question arises, however, how in an age of unparalleled violence, when, as Gandhi points out, 'the atom bomb has deadened the finest feelings that sustained mankind for ages',[460] does one go about the process of reconnecting with the laws of truth within oneself? How do we actually go about the process of resensitization?

The answer to this is as varied as the persons willing to commit themselves to the Path. It must be this way, as the *question* of balance can only be answered by *inner* balance, and this is ultimately intensely personal. Nonviolence, like the soft energy paths, is not conducive to monolithic or centralized definition. It is rather individual and must be appropriated to the facts of any given situation.

There are, however, certain principles that can be suggested as guidelines:

1. Become aware of the problem. Just knowing the full dimensions of the nuclear weapons/reactor complex in society and the degree each of us is affected by it, is half the battle won, for realizing the depth and extent of the problem will yield of its own accord a way into becoming part of the solution.

2. Once the dimensions of the problem are realized, there should arise a personal commitment not to co-operate with its evil. This means negatively the eschewing of personal violence as well as societal violence. Put positively, this non-co-operation involves active participation with alternative energetics and alternative ways of relating with oneself and other people. This may involve insulating one's home, installing a solar panelling system or conserving energies in existing systems; it may involve writing to your government representative about your concerns, actively joining in protest marches and occupations of nuclear weapons/reactor sites or merely informing friends of the problem; and it may involve different practices available that help you re-connect with your own centre, such as yoga, meditation or prayer,

3. Fundamental to nonviolence is a refusal to divide the world into 'us' and 'them'. Violence only exists where there is dichotomy between people. It is imperative, therefore, that no matter how much you struggle to end nuclear power, it be always kept in mind that the radiation which effects our bodies and mutates our children damages the body of the utility company president and his children as well. It makes no difference whether one is a police officer, an atomic worker, a demonstrator or a student; radiation is dangerous to all. In an age of overkill, all are equally threatened and therefore all must be brought together against the common threat.

Robert Lifton makes the point that the scientists involved in the construction of the bombs dropped on Hiroshima and Nagasaki fell into three categories as they reflected upon what they had done. One group, symbolized most powerfully by Edward Teller, who insisted after the war on the construction of the hydrogen bomb, fell into the psychological state of *identification* with the bomb, embracing it as a scientific advance that would greatly enhance national security. Another group, exemplified by such men as Niels Bohr and Leo Szilard, manifested the opposite psychological state, that of *survivor mission*, opposing nuclear weapons and seeking meaningful disarmament and world peace. The majority of the scientists remained between these two extremes, continuing to do what they were told to do in terms of continuing to research and develop nuclear weapons and yet being filled with gnawing anxiety and self-doubt as to the morality of what they were doing.

It is within one of these three categories that each one of us finds ourself. All of us are impacted by nuclear power and therefore all must choose whether through psychic numbing to its devastation we can come to identify with it; whether through psychic opening-up we can become part of the movement whose mission it is to direct history away from the hard path to the soft path and from an age of overkill to an age of planetary co-operation; or whether we will remain psychologically ambivalent and divided, aware of the issue but because of a lack of will or a dependence upon the system paralysed to act.

Human history will be either terminated, constructively guided, or continued but in a damaged, possibly mutated form, according to which of these choices we collectively make.

NOTES

1. For a complete account of scientific research under the Nazis, see David Irving, *The Virus House: Germany's Atomic Research and Allied Countermeasures,* London 1967, passim; also Arthur Compton, *Atomic Quest, A Personal Narrative,* London 1956, pp. 1-64; William Manchester, *The Glory and the Dream: A Narrative History of America, 1932-1972,* London 1975, pp. 335-38. For an account of Japanese research, see Pacific War Research Society, *The Day Man Lost: Hiroshima, 6 August, 1945,* Tokyo 1972, Part I.

2. Alice Kimball Smith, *A Peril and a Hope: The Scientists' Movement in America, 1945-47,* New York/London 1965, p. 3; also Henry L. Stimson, 'The Decision to Use the Atomic Bomb', *Harper's Magazine,* Vol. 194, No. 1161, New York/London, February 1947, p. 98. For account of the earliest experiments from one involved in them, see Edward Teller with Allen Brown, *The Legacy of Hiroshima: The Atomic Story,* New York 1962, pp. 7-10; also John Campbell, *The Atomic Story,* New York 1947, pp. 3-116.

3. Gordon Thomas and Max Morgan-Witts, *Ruin from the Air: The Atomic Mission to Hiroshima,* London 1977, pp. 6f; William L. Lawrence, *Dawn over Zero: The Story of the Atomic Bomb,* London 1947, p. 71.

4. Stimson, 'Decision to Use the Atom Bomb', p. 98; for account of early US research and development from the military commander in charge of it, see Leslie R. Groves, *Now It Can Be Told: The Story of the Manhattan Project,* London 1963, Part I; also Lawrence, *Dawn over Zero,* pp. 69-145. For account of British research and development, see Ronald W. Clark, *The Birth of The Bomb: The Untold Story of Britain's Part in the Weapon that Changed the World,* London 1961, pp. 1-158.

5. James F. Byrnes, *Speaking Frankly,* London 1947, p.257; also Smith, *A Peril and a Hope,* p. 4; Stimson, '*Decision to Use. . .*', p.98.

6. For full account of the history of the Manhattan Engineer Project, see Stephanie Groueff, *Manhattan Project: The Untold Story of the Making of the Atomic Bomb,* London 1967, passim; also Groves, *Now It Can Be Told,* pp. 3-18; Walter Smith Schoenberger, *Decision of Destiny,* Columbus, Ohio 1969, pp. 18-57. For particular emphasis on scientific problems encountered, see Campbell, *The Atomic Story,* pp. 117-227; also E. Teller and Albert L. Lutter, *Our Nuclear Future. . .Fact, Dangers and Opportunities,* London 1958, passim; Compton, *Atomic Quest,* pp. 65-145.

7. Thomas and Morgan-Witts, *Ruin from the Air*, p. 8.
8. Schoenberger, *Decision of Destiny*, pp. 24-6.
9. Teller and Brown, *Legacy of Hiroshima*, p. 24.
10. Harry S. Truman, *Year of Decisions: 1945*, London 1955, p. 10.
11. Ibid,. p. 11; Burt Hirshfeld, *A Cloud Over Hiroshima: The Story of the Atomic Bomb*, Folkestone 1974, pp. 87-9.
12. Henry L. Stimson, *On Active Service in Peace and War*, London 1947, p. 376.
13. Schoenberger, op. cit., pp. 116ff.; also Frank Chinnock, *We of Nagasaki: The Forgotten Bomb*, London 1970, pp. 35f.
14. 'Decision to Use. . .', pp. 99f. for full text of memo.
15. Stimson, *On Active Service*, p. 363.
16. Len Giovannitti and Fred Freed, *The Decision to Drop the Bomb*, London 1967, p. 56. For more on the Interim Committee, see Herbert Feis, *The Atomic Bomb and the End of World War II*, Princeton 1968, pp. 38-40, 42, 50, 51n; Schoenberger, op.cit., pp. 116-59; Stimson, *On Active Service*, pp. 363ff.; Richard G. Hewlett and Oscar E. Anderson Jr, *The History of the United States Atomic Energy Commission*, Vol. 1, *The New World, 1939 /1946*, University Park, Pa. 1962, pp. 353-61.
17. On the Scientific Panel, see Feis, *The Atomic Bomb*, pp. 39-57, 171, 196; Hewlett and Anderson, *History of the US Atomic Energy Commission*, pp. 345f.
18. For a discussion of these recommendations by two men on the committee, see Stimson, 'Decision to Use. . .', pp. 100f.; Byrnes, op.cit., p. 261. For commentary: Feis, op.cit., pp. 38-57; Smith, *A Peril and a Hope*, pp. 34-43, 48-54; and Giovannitti and Freed, op.cit., 53-68. For critical analysis by a scientist involved in the project, Teller, *Legacy*, pp. 13-15.
19. 'Decision to Use. . .', p.101.
20. Ibid.
21. Giovannitti and Freed, *The Decision to Drop the Bomb*, p. 134; Thomas and Morgan-Witts, *Ruin from the Air*, p. 160; Stimson, 'Decision to Use. . .', p. 102; Byrnes, *Speaking Frankly*, pp. 261f.
22. Giovannitti and Freed, op.cit., p. 31.
23. Feis, *The Atomic Bomb and the End of World War II*, p. 8.
24. Winston Churchill, *The Second World War, Vol. VI, Triumph and Tragedy*, London 1954, p. 522.
25. Giovannitti and Freed, op.cit., p. 90.
26. Feis, op.cit., p. 192; Churchill, op.cit., p. 542; Thomas and Morgan-Witts, *Ruin*, pp. 50ff. For cultural study of this trait, Kurt Singer, *Mirror, Sword and Jewel: A Study of Japanese Characteristics*, London 1973, pp. 149-168.
27. United States Strategic Bombing Survey, Report 62, 'Japanese Air Power', in Feis, p. 193.
28. Giovannitti and Freed, op.cit., pp. 87f.
29. In *'Harper's Magazine'*, Vol. 194, No. 1161, New York/London, February 1947, p.106.
30. Stimson, *On Active Service*, pp. 372, 373; for Churchill's con-

currence, *Triumph and Tragedy*, p. 522. See also Schoenberger, *Decision of Destiny*, pp. 304-307.

31. Stimson, 'Decision to Use. . .', pp. 106, 107 (emphasis mine); also Groves, *Now It Can Be Told*, p. 319; Robert Butow, *Japan's Decision to Surrender*, Stanford 1954, p. 180.

32. Stimson, op.cit., pp. 106, 107.

33. Ibid., p. 106.

34. Churchill, *Triumph and Tragedy*, p. 552.

35. Feis, op.cit., p. 191.

36. Giovannitti and Freed, op.cit., pp. 34f.; Thomas and Morgan-Witts, *Ruin from the Air*, pp. 103-105, 107f.

37. Ibid., p. 35. For further documentation of the extent of US devastation of Japan, see David H. James, *The Rise and Fall of the Japanese Empire*, London 1951, pp. 320-22; also Charles Buteson, *The War With Japan: A Concise History*, London 1968, pp.375f.

38. Giovannitti and Freed, *The Decision to Drop the Bomb*, p. 42.

39. Dwight D. Eisenhower, *The White House Years: Mandate for Change, 1953-56*, New York 1963, pp. 312-13.

40. Hewlett and Anderson, *A new World*, pp. 357-58, 360f.

41. P. Kecskemeti, *Strategic Surrender: The Politics of Victory and Defeat*, Stanford 1958, p. 196.

42. Feis, *Atomic Bomb*, pp. 179-89.

43. Kecskemeti, *Strategic Surrender*, pp. 198ff.

44. G. Alperovitz, *Atomic Diplomacy: Hiroshima and Potsdam: The Use of the Atomic Bomb and the American Confrontation With Soviet Power*, London 1966, p. 238.

45. Byrnes, *Speaking Frankly*, p. 92.

46. Alperovitz, *Atomic Diplomacy*, p. 237.

47. Hewlett and Anderson, *A New World*, p. 357.

48. Giovannitti and Freed, *The Decision to Drop the Bomb*, p. 42.

49. Ibid.

50. Feis, *Atomic Bomb*, p. 188.

51. See ibid.

52. Feis, op.cit., p. 191 (emphasis mine).

53. Alperovitz, *Atomic Diplomacy*, p. 239.

54. Ibid., p. 227.

55. Leahy, *I was There*, New York 1950, p. 429.

56. Alperovitz, *Atomic Diplomacy*, p. 227.

57. H. Feis, *Between War and Peace: The Potsdam Conference*, Princeton/London 1960, p. 171.

58. Churchill, *Triumph and Tragedy*, pp. 68, 76.

59. Ibid., p. 238.

60. Alperovitz, *Atomic Diplomacy*, p. 228.

61. C.E. Black, 'Soviet Policy in Eastern Europe', in *Annals of the American Academy of Political and Social Science*, Vol. 263, Philadelphia, May 1949, p. 155.

62. R. Dennet and J.E. Johnson, (eds), *Negotiating with the Russians*, Boston 1951, p. 178.

63. In Black, 'Soviet Policy', p. 155.
64. Alperovitz, *Atomic Diplomacy*, p. 229.
65. In A. Bryant, *Triumph in the West 1943-1946*, London 1959, pp. 447, 478.
66. Truman, *Year of Decisions*, p. 90.
67. Alperovitz, *Atomic Diplomacy*, pp. 13, 230f.
68. Truman, op.cit., p. 90.
69. Alperovitz, op.cit., p. 231. Contra Feis, who argues that due to American secrecy, 'the light of the explosion brighter than a thousand suns filtered into the conference rooms at Potsdam only as a distant gleam'. *Between War and Peace*, p. 180.
70. Alperovitz, op.cit., p. 232.
71. *War Memories*, Vol. III, *Salvation*, New York/London 1960, p. 230.
72. Churchill, *Triumph and Tragedy*, p. 553.
73. Margaret Gowing, *Britain and Atomic Energy*, 1939-45, London 1964, p. 341.
74. Ibid.
75. Giovannitti and Freed, *The Decision to Drop the Bomb*, p. 58. See also Gowing, op.cit., pp. 346-55 for details of the meeting. For full account of the entirety of Bohr's efforts with Roosevelt and Truman as well as Churchill, see Smith, *A Peril and a Hope*, pp. 5-14; also Clark, *The Birth of the Bomb*, pp. 176-80.
76. Gowing, op.cit., p. 371.
77. Giovannitti and Freed, op.cit., pp. 63-7.
78. In Smith, op.cit., p. 44. For full text of report, Smith, pp. 560-72. For discussion, Feis, *Atomic Bomb*, p. 52; Smith, pp. 41-8.
79. Feis, *Atomic Bomb*, p. 54.
80. Smith, *A Peril and A Hope*, p. 49.
81. Ibid, p. 54.
82. Even the survivors of Hiroshima recognized the fact that had Japan developed the bomb first it would have dropped it. See Robert J. Lifton, *Death in Life: Survivors of Hiroshima*, New York 1967, p. 325.
83. William Lawrence, *Dawn Over Zero: The Story of the Atomic Bomb*, London 1947, pp. 10f. For details of testing procedures and results, see Giovannitti and Freed, *Decision to Drop the Bomb*, pp. 194-98, 238; Thomas and Morgan-Witts, pp. 202-206.
84. In Giovannitti and Freed, *The Decision to Drop the Bomb*, p. 197; see also Oppenheimer, *The Flying Trapeze: Three Crises for Physicists*, London 1964, p. 59.
85. In Lawrence, *Dawn Over Zero*, p. 9.
86. In Thomas and Morgan-Witts, *Ruin from the Air*, p. 206. For memo Groves sent to Stimson at Potsdam, see Feis, *Between War and Peace*, pp. 165-71. For Groves' later evaluation, *Now it Can be Told*, pp. 298f.
87. Thomas and Morgan-Witts, op.cit., p. 209.
88. Truman, *Year of Decisions*, p. 345.
89. Feis, *Atomic Bomb*, p. 87.
90. Ibid.

91. Manchester, *The Glory and the Dream*, p. 379.

92. Stimson, 'Decision to Use. . .', pp. 101-5. For detailed background of Japanese rejection, see Butow, *Japan's Decision to Surrender*, pp. 142-50.

93. 'Decision to Use. . .', p. 105. For further discussion of target selection critieria, see Groves, *Now It Can Be Told*, pp. 263-76, 316, 343; Feis, *Atomic Bomb*, p. 85; Giovannitti and Freed, *The Decision to Drop the Bomb*, pp. 38f; Thomas and Morgan-Witts, op.cit., pp. 155f.

94. For detailed check-list as to what this involved, see Thomas and Morgan-Witts, op.cit., pp. 304f.

95. Feis, *Atomic Bomb;* Thomas and Morgan-Witts, op.cit., p. 305.

96. Thomas and Morgan-Witts, op.cit., p. 312.

97. W.F. Craven and J.L.Cate, (eds), *The Army Air Forces in World War II*, Vol. IV, *The Pacific – Matterhorn to Nagasaki*, Chicago/London 1953, p. 716.

98. D.R. Inglis, 'The Nature of Nuclear Warfare', *Nuclear Weapons and the Conflict of Conscience*, ed. John C. Bennett, London 1962, p. 42.

99. F. Clune, *Ashes of Hiroshima*, London 1950, p. 90.

100. Lifton, *Death in Life*, pp. 15-18.

101. Feis, *Atomic Bomb*, p. 48; also 50, 75; also Smith, *A Peril and a Hope*, pp. 3-72.

102. Lifton, op.cit., p. 18. For similar description, see Pacific War Research Society, *The Day Man Lost*, pp. 231-36; Chinnock, *We of Nagasaki*, pp. 71-85, 102.

103. In *Children of the A-Bomb: The Testament of the Boys and Girls of Hiroshima*, ed. Arata Osaka, tr. Jean Dan and Ruth Sieben-Morgan, London 1959, pp. 129, 130.

104. Thomas and Morgan-Witts, *Ruin from the Air*, pp. 326, 324. For other reactions from those observing from the air, Lawrence, *Dawn over Zero*, p. 179ff.

105. Dr M. Hachiya, *Hiroshima Diary*, tr. and ed. Warner Wells, Chapel Hill, NC 1955, p. 69.

106. Lifton, op.cit., p. 19.

107. Craven and Cate, *The Pacific*, pp. 722-23; Lawrence, op.cit., p. 209; Lifton, *Death in Life*, p. 20.

108. Ashley Oughterson and Shields Warren, *Medical Effects of the Atomic Bomb on Japan*, New York 1956, passim.

109. S. Nagaoka, *Hiroshima Under Atomic Attack*, Hiroshima, no date given.

110. Lifton, *Death in Life*, p. 20, 509 n. 6.

111. *The Day Man Lost*, pp. 244-53; for breakdown of deaths in civilian and military categories, pp. 254-65, 266-76 respectively.

112. Lifton, op.cit., p. 29; also Hachiya, *Hiroshima Diary*, p. 21, who comments: 'Hiroshima was no longer a city but a burnt-over prairie.'

113. Lifton, op.cit., p. 79.

114. Feis, *Atomic Bomb*, p. 123.

115. Butow, *Japan's Decision to Surrender*, p. 152 n. 35.

116. Feis, *Atomic Bomb*, J. Toland, *The Rising Sun: The Decline and Fall*

of the Japanese Empire: 1936-1945, London 1970, pp. 793-99.

117. Giovannitti and Freed, *The Decision to Drop the Bomb*, p. 265.

118. Ibid., p. 269.

119. Chinnock, *We of Nagasaki*, p. 90; for description from the bomber crew, pp. 123-26. See also Craven and Cate, Vol. V, p. 719ff.; Feis, *Atomic Bomb*, 128ff; Groves, *Now It Can Be Told*, pp. 341-55; Toland, op.cit., pp. 793-809. For medical details, Oughterson and Warren, op.cit., esp. pp. 1-5, 97-190. For personal recount by survivors, Richard L. Gage, (ed.), *Cries For Peace*, Tokyo 1978, pp. 195-234.

120. .Chinnock, op.cit., p. 286.

121. Feis, *Atomic Bomb*, pp. 131f., 139f., 141; Toland, pp; 810ff.; Rikihei Inogushi, Tadashi Nakajima and Roger Pineau, *The Divine Flame: Japan's Kamakaze Force in World War II*, London 1959, pp. 160f.

122. Lifton, *Death in Life*, p. 115.

123. Other accounts: M. Hachiya, *Hiroshima Diary;* T. Nagai, *We of Nagasaki*, New York 1951; H. Agawa, *Devil's Heritage*, Tokyo 1957; Arata Osada, *Children of the A-Bomb;* Robert Jungk, *Children of Ashes*, New York 1961; John Hersey, *Hiroshima*, New York 1959; Robert Trumbull, *Nine Who Survived Hiroshima and Nagasaki*, Tokyo/Rutland, VT. 1957.

124. Lifton, op.cit., p. 479.

125. Ibid.

126. Pedro Arrupe, *A Planet to Heal*, Rome 1975, p. 26.

127. Lifton, op.cit., p. 21.

128. Ibid., p. 479.

129. Gage, *Cries for Peace*, p. 154. For chronicle of 6 August as it happened hour by hour, see Pacific War Research Society, *The Day Man Lost*, Part III.

130. Lifton, op.cit., p. 22.

131. Ibid.

132. Ibid., p. 23; also p. 38.

133. Hachiya, *Hiroshima Diary*, p. 21.

134. Ibid., p. 16; Pacific War Research Society, *The Day Man Lost*, p. 241.

135. Pacific War Research Society, op.cit., p. 242.

136. Hachiya, *Hiroshima Diary*, p. 70.

137. Lifton, *Death in Life*, p. 25.

138. Ibid., p. 27.

139. Gage, *Cries for Peace*, p. 162.

140. *Hiroshima Diary*, p. 27.

141. *Death in Life*, p. 256.

142. Ibid., p. 30.

143. See Erick H. Erickson, 'The Problem of Ego Identity', *Journal of the American Psychoanalytic Association*, Vol. 3, 1955, pp. 447-66; also Martin Grotjahn, 'Ego Identity and the Fear of Death and Dying', *Journal of the Hillside Hospital*, Vol. 9, 1960, pp. 147-55.

144. *Death in Life*, p. 30.

145. *Hiroshima Diary*, passim, but especially pp. 72f., 178, 179.

146. Lifton, *Death in Life*, p. 77.

147. Ibid., p. 57.

148. Ibid., p. 59.

149. Ibid., p. 60.

150. The most extensive studies of the delayed after-effects of the Hiroshima and Nagasaki bombings were done under the sponsership of the Atomic Bomb Casualty Commission, published in 'Medical Findings and Methodology of Studies by the Atomic Bomb Casualty Commission on Atomic Bomb Survivors in Hiroshima and Nagasaki', in *The Use of Vital and Health Statistics for Genetic and Radiation Studies, Proceedings* of a seminar sponsored by the United Nations and the World Health Organization, Geneva, September 1960. See also J.W. Hollingsworth, 'Delayed Radiation Effects in Survivors of the Atomic Bombings', *New England Journal of Medicine*, Vol. 263, 1960, pp. 381-487; Frank Barnaby, 'The Continuing Body Count at Hiroshima and Nagasaki', *The Bulletin of the Atomic Scientists*, Vol. 33, No. 10, December, 1977, pp. 48-53.

151. Lifton, op.cit., p. 104; see also Edward L. Socolow, 'Thyroid Carcinoma in Man After Exposure to Ionizing Radiation: Summary of the Findings in Hiroshima and Nagasaki', *New England Journal of Medicine*, Vol. 268, 1963, pp. 406-10.

152. The most extensive work on the genetic problem has been done by James V. Neal and W.O. Schull, 'Radiation and Sex Ratio in Man: Sex Ratio among Children of Atomic Bombings Suggests Induced Sex-Linked Lethal Mutations', *Science*, Vol. 128, 1958, pp. 343-48; also *The Effect of Exposure to the Atomic Bomb on Pregnancy Termination in Hiroshima and Nagasaki*, Washington, DC 1956.

153. For account of one hibakusha family involved in giving birth to a microcephalic child, see Gage, *Cries for Peace*, pp. 181-83.

154. Lifton, *Death in Life*, pp. 105f.; 116f.

155. Ibid., p. 119.

156. Ibid., p. 225.

157. Ibid., p. 482.

158. See Erick Lindermann, 'Symptatology and Management of Acute Grief', *American Journal of Psychiatry*, Vol. 101, 1944, pp. 141-48.

159. In Lifton, op.cit., p. 484.

160. Karl Stern, 'Death Within Life', *Review of Existential Psychology and Psychiatry*, Vol. 2. 1962, pp. 141-44.

161. Lifton, op.cit., p. 484.

162. Ibid.

163. Ibid., p. 486.

164. Ibid.

165. Niederland, 'Psychiatric Disorders Among Persecution Victims', *Journal of Nervous and Mental Disease*, Vol. 139, 1964, pp. 458-74, 468.

166. 'Thoughts for the Times on War and Death', tr. A.A. Brill and A.B. Kutler, in *The Standard Edition of the Complete Psychological Works of Sigmund Freud*, ed. James Strachey, London 1957, Vol. XIV, p. 293.

167. Elie Wiesel, *Night*, New York 1960, p. 92.

168, Lifton, *Death in Life*, p. 496.
169, Ibid., p. 36.
170. See Joan M. Erickson, 'Eye to Eye', *Man Made Object*, ed. Gyorgy Kepes, New York 1966.
171. Lifton, op.cit., p. 499.
172. Niederland, 'Psychiatric Disorders', p. 463.
173. *Night*, p. 45.
174. S. Rado, 'Psychodynamics and Depression from the Etiological Point of View', in his *Psychoanalysis of Behavior*, New York 1956, p. 238.
175. E. Luby, 'An Overview of Psychosomatic Diseases', *Psychosomatics*, Vol. 4, 1963, pp. 1-8, 7.
176. Bastiaans in Lifton, op.cit., pp. 503, 504.
177. Bruno Bettleheim, *An Informed Heart*, Glencoe, Ill, 1960, p. 261.
178. Lifton, op.cit., pp. 511ff.
179, Hachiya, *Hiroshima Diary*, p. 63. Cf. with Wards's response at announcement of the Emperor's surrender, pp. 98ff.
180. Lifton, op.cit., p. 80.
181. Ibid., pp. 80, 81.
182. Bettleheim, *An Informed Heart*, pp. 171, 172.
183. Lifton, op.cit., p. 511.
184. Ibid., p. 514.
185. Freud, 'Thoughts for the Times on War and Death', p. 289.
186. Elias Canetti, *Crowds and Power*, New York 1962, pp. 227, 230, 448; also Lifton op.cit., p. 521.
187. In Lifton, op.cit., p. 392.
188. Ibid.
189. Ibid., p. 387.
190. Ibid., p. 389.
191. Ibid., p. 393.
192. Ibid.
193. Hachiya, *Hiroshima Diary*, p. 38.
194. Lifton, *Death in Life*, p. 369.
195. Ibid.
196. Ibid., pp. 373, 374. Also Gage, *Cries for Peace*, pp. 184-86.
197. Ibid., pp. 372, 373.
198. Ibid., p. 382.
199. Freud, *Moses and Monotheism*, New York 1955, p. 139.
200. E. Erikson, *Young Man Luther*, New York and London 1958, p. 252.
201. Lifton, op.cit., 211, 212.
202. Joseph Campbell, *The Hero With a Thousand Faces*, New York 1956, passim.
203. Lifton, op.cit., p. 302.
204. Ibid.
205. Ibid., pp. 304, 305.
206. Ibid., p. 304.
207. H. Feis, *From Trust to Terror: The Onset of the Cold War, 1945-1950*, London 1970, p. 91.

208. Teller, *Legacy of Hiroshima*, p. 24.
209. Feis, *From Trust to Terror*, p. 93.
210. Truman, *Year of Decisions*, pp. 472, 473.
211. *Public Papers of the Presidents of the United States*, Vol. 1, *April 12-December 31, 1945*, Washington, DC 1961, p. 1.
212. Feis, *From Trust to Terror*, pp. 170, 171.
213. Ibid, p. 171.
214. Stimson, *On Active Service*, pp. 376-81.
215. For a thorough discussion of the movement of scientists toward international control of atomic energy during the period of time in question, August-September, see Smith, *A Peril and a Hope*, pp. 75-127.
216. Byrnes, *Speaking Frankly*, pp. 277-97.
217. Stimson, *On Active Service*, pp. 382f.
218. Ibid, p. 383n.
219. Truman, *Year of Decisions*, p. 463; also, Alperovitz, *Atomic Diplomacy*, 195-200.
220. For a detailed account of the meeting, see Hewlett and Anderson, pp. 418-21; also, Truman, *Year of Decisions*, pp. 463-65.
221. Truman, ibid., p. 465.
222. Ibid., p. 466.
223. For full text of statement, ibid., pp. 481-83.
224. Truman, *Years of Trial and Hope, 1964-1953*, London 1956, p. 11; also Feis, *From Trust to Terror*, pp. 140-43.
225. Feis, ibid., pp. 143-45.
226. Ibid., p. 140; also Hewlett and Anderson, *A New World*, pp. 531-619.
227. Feis, ibid., p. 400.
228. Truman, *Year of Decisions*, p. 483.
229. In Feis, *From Trust to Terror*, p. 401.
230. Ibid., p. 407, (emphasis mine).
231. Lerner, *The Age of Overkill: A Preface to World Politics*, London 1964, pp. 34ff.
232. Feis, *From Trust to Terror*, p. 412.
233. In John Kenneth Galbraith, 'Age of Uncertainty' a documentary for BBC television, 3 April 1977.
234. Teller, *Legacy of Hiroshima*, p. 43.
235. Ibid., pp. 43, 44.
236. Ibid., pp. 27-33. For account of the political milieu in general in which Truman made this decision, see Alan D. Harper, *The Politics of Loyalty: The White House and the Communist Issue, 1946-1952*, Westport, CT 1969, passim.
237. For account of the hydrogen bomb programme from the 'father of the H-bomb', see Teller, *Legacy of Hiroshima*, pp. 34-57; for critical assessment, see Hans Bethe, 'The American Hydrogen Bomb', in *The Atomic Age: Scientists in National and World Affairs: Articles from 'The Bulletin of the Atomic Scientists', 1945-1962*, ed. Morton Grodzins and Eugene Rabinovitch, New York/London 1963, pp. 144-62. See also Ralph Lapp, *Kill and Overkill: The Strategy of Annihilation*, London 1962,

pp. 29ff.

238. Lapp, op.cit., p. 44.

239. *World Armaments and Disarmaments,* London 1978, pp. 423-25. See also Herbert York, 'The Nuclear "Balance of Terror" in Europe', *Bulletin of the Atomic Scientists,* Vol. 32, No. 5, Chicago May, 1976, pp. 9-17; The International Institute for Strategic Studies, *The Military Balance: 1977-1978,* London 1977; R.T. Pretty, ed., *Janes Weapon System, 1976,* London 1977; J.I. Coffray, *Arms Control and European Security: A Guide to East-West Relations,* London 1977.

240. *International Herald Tribune,* 7 June 1977, p. 1.

241. Ibid.

242. Ibid.

243. 1978 Yearbook, *World Armaments and Disarmaments,* p. 6.

244. Ibid.

245. Ibid.

246. *International Herald Tribune,* Tuesday 29 May 1979, p.1.

247. *Newsweek,* 21 May 1979, p. 13.

248. In W.C. Patterson, *Nuclear Power,* London 1976, p. 133.

249. Ibid, p. 139.

250. *Honicker v. Hendrie: A Lawsuit to End Atomic Power,* Summertown, TN 1978, p. 25, (hereafter: *HVH*).

251. Patterson, *Nuclear Power,* pp. 148f.

252. In *Honicker v. Hendrie, (HVH),* p. 29.

253. *International Herald Tribune,* Monday 11 June 1979, p. 3.

254. *HVH,* p. 30.

255. In John Cox, *Overkill: The Story of Modern Weapons,* Middlesex/ New York 1977, p. 79.

256. H. Kissenger, 'Central Issues in American Foreign Policy', *Agenda for the Nation,* ed. K. Gordan, New York 1968, p. 501.

257. York, 'The Nuclear "Balance of Terror" in Europe', p. 10.

258. Michael Klare, 'The Political Economy of Arms Sales', *Bulletin of the Atomic Scientists,* Vol. 32, No. 9, Chicago, November 1976, p. 11.

259. Ibid.

260. See the Stockholm International Peace Research Institute, 1978 Yearbook, *World Armaments and Disarmaments,* for a country-by-country analysis or arms sales, pp. 133-65.

261. In Klare, op.cit., p. 17.

262. Ibid., p. 18.

263. In Elaine Davenport, Paul Eddy and Peter Gillman, *The Plumbat Affair,* London 1978, pp. 175, 176.

264. Ibid.

265. Howard Kohn and Barbara Newman, 'How Israel Got the Nuclear Bomb', *Rollingstone,* 1 December 1977, pp. 38-40.

266. Davenport et al., *The Plumbat Affair,* passim, esp., pp. 14-110.

267. Kohn and Newman, op.cit., pp. 38-40; Davenport et al., op.cit., p. 175.

268. *New York Times,* Thursday 2 March 1978.

269. Davenport et al., op.cit., p. 174.

270. Paul Levanthal quoted in Robert Jungk, *The Nuclear State*, tr. Eric Mosbacher, London 1979, pp. 91, 92.
271. Ibid., p. 95.
272. Ibid., p. 92.
273. John Berger, *Nuclear Power: The Unviable Option: A Critical Look at Our Energy Alternatives*, Palo Alto, CA 1976, p. 220.
274. *Foreign Policy*, New York, Summer 1976, in Jungk, *The Nuclear State*, p. 100.
275. *Machete*, April 1973, in *Nuclear State*, p. 100.
276. *Nuclear State*, p. 92.
277. 'Nuclear Journal Report', 22 March 1975, in *Nuclear State*, p. 93.
278. Stockholm International Peace Research Institute, 1978 Yearbook, *World Armaments and Disarmaments*, p. 133.
279. Abdus Salam, 'Ideals and Realities', *Bulletin of the Atomic Scientists*, Vol. 32, No. 7, September 1976, p. 15.
280. In Klare, 'The Political Economy of Arms Sales', p. 12.
281. Salam, 'Ideals and Realities', p. 10.
282. Herman Kahn, *Thinking the Unthinkable*. London 1962, p. 18.
283. Ibid, pp. 18f.
284. Lerner, *Age of Overkill*, p. 47.
285. L. Dumas, 'National Security in the Nuclear Age', *Bulletin of the Atomic Scientists*, Vol. 32, No. 4, May 1976, pp. 24-35.
286. Milton Leitenberg, 'Accidents of Nuclear Weapons and Nuclear Weapons Delivery Systems' in the Stockholm International Peace Research Institute, *1968 Yearbook*, London 1969, pp. 259-70.
287. J.B. Phelps, *Accidental War: Some Dangers in the Sixties*, Columbus, Ohio 1960, p. 8.
288. Lapp, *Kill and Overkill*, p. 127; Dumas, 'National Security in the Nuclear Age', p. 27.
289. J. Larus, *Nuclear Weapons Safety and the Common Defense*, Columbus, Ohio 1967, pp. 94-9; Dumas, op.cit., p. 27.
290. Dumas, op.cit., p. 27.
291. Ibid.
292. Ibid., p. 28.
293. Ibid.
294. *The Atomic Energy Commission Dictionary*, Washington, DC 1971, p. 31.
295. D.L. Crouson, 'Safeguards and Nuclear Materials Management in the U.S.A.', Atomic Energy Commission document, Washington, DC 1970, p. 8.
296. Dumas, 'National Security', p. 31.
297. Symington, *New York Times*, 11 November 1970, p. 1.
298. Ibid., 22 September 1974, p. 3.
299. In D. Middleton, 'Could a U.S. Atom Bomb be Stolen' in ibid.
300. Dumas, op.cit., p. 32.
301. Ibid.
302. Jerome B. Weisner and Herbert F. York, 'National Security and the Nuclear Test Ban', *Scientific American*, October 1964, p. 32.

303. Richard Garwin, 'Anti-Submarine Warfare and National Security', *Scientific American*, July 1972, p. 19.
304. H.Scoville, 'Missile Submarines and National Security', *Scientific American*, June 1972, p. 23.
305. Dumas, 'National Security', p. 35.
306. Jonathan Steele article, *The Guardian*, 9 October 1975.
307. Ibid; also Cox, *Overkill*, pp. 124f.
308. York, 'The "Balance of Terror" in Europe', p. 13; based on S. Glasstone, *The Effects of Nuclear Weapons*, Washington DC 1962, and Arthur Westing to Herbert York, August 1975.
309. Ibid, p. 16. See also pp. 12-14.
310. Ibid, p. 11. Adapted from Jeffrey Record, *U.S. Nuclear Weapons in Europe*, Washington DC 1974.
311. *International Herald Tribune*, 16 May 1979, p. 1.
312. See above, pp. 52-3.
313. Lerner, *Age of Overkill*, pp. 20-23.
314. Ibid., p. 22.
315. Robert Ardrey, *African Genesis*, London 1961, p. 316.
316. Ibid., p. 317.
317. Ibid., p. 26.
318. Ibid., p. 324.
319. Robert Bigelow, *The Dawn Warriors: Man's Evolution Towards Peace*, London 1969, p. 3.
320. Ibid., p. 4.
321. Ibid.
322. Lerner, *Age of Overkill*, p. 26.
323. Berger, *Nuclear Power: The Unviable Option*, p. 100.
324. Ibid., p. 75.
325. Quoted by Claire Ryle et al., 'Radiation Risk', *Vole*, 3 October 1979, p. 10.
326. Berger, op.cit., p. 45.
327. Patterson, *Nuclear Power*, pp. 42-86.
328. Berger, op.cit., p. 45.
329. U.S. Atomic Energy Commission, Washington, DC March 1957; see also Berger, p. 53; Patterson, pp. 170ff.
330. Berger, op.cit., p. 55.
331. Ibid., p. 54.
332. Ibid., pp. 54, 55.
333. 'Reactor Safety Study: An Assessment of Accident Risks in U.S. Commercial Nuclear Power Plants', WASH 1400/NUREG 75/014, Washington, DC, October 1975. For critical analysis, see Berger, op.cit., pp. 63-72.
334. Berger, ibid., p. 64.
335. For other details, see Patterson, *Nuclear Power*, pp. 175f.
336. Ibid., pp. 162-66.
337. Ibid., p. 163.
338. Alan Roberts and Zhores Medvedev, *Hazards of Nuclear Power*, Nottingham 1977, pp. 58-70; see also Medvedev, 'Two Decades of Dis-

sidence', *New Scientist*, Vol. 72, 30 June 1977, p. 264.

339. Patterson, op.cit., p. 161.
340. Ibid., pp. 185-88.
341. Ibid., pp. 198, 199.
342. Ibid., pp. 210-12.
343. *HVH*, p. 83.
344. Ibid.
345. Patterson, op.cit., p. 187.
346. Roberts and Medvedev, *Hazards of Nucelar Power*, pp. 58f.
347. *HVH*, p. 82.
348. Ibid., p. 83.
349. Based on *Time*, 9 April 1979, p. 12.
350. *Washington Post*, 8 April, 1979, p. 1.
351. *Washington Post*, 3 April 1979, p. 1.
352. Ibid.
353. Based on *Newsweek*, 9 April 1979, p. 11.
354. *Washington Post*, 8 April 1979, p. 1.
355. Ibid.
356. Based on *Newsweek*, 9 April 1979. p. 9.
357. Ibid.
358. Ibid., 9 April 1979, p. 1.
359. Ibid.
360. Ibid.
361. Ibid.
362. K. Morgan, 'Cancer and Low Level Ionizing Radiation', *Bulletin of the Atomic Scientists*, Vol. 34, No. 7, September 1978, pp. 30, 31.
363. Patterson, *Nuclear Power*, p. 30; also pp. 280-285 for detailed analysis.
364. Morgan, 'Cancer', p. 32. For additional information see also p.21.
365. *HVH*, p. 17.
366. Tom Barry. 'Bury My Lungs at Red Rock', *The Progressive*, February 1979, p. 26.
367. Ibid., p. 27.
368. Ibid.
369. Howard Kohn, 'Malignant Giant', *Rollingstone*, 27 March 1975, p. 18.
370. *HVH*, p. 17.
371. For more information on this, *HVH*, pp. 17-19, 38, 39; also Patterson, *Nuclear Power*, pp. 87-90.
372. *HVH*, p. 39.
373. Ibid., p. 42.
374. Ibid.
375. Barry, 'Bury My Lungs at Red Rock', p. 26. For more on milling, see *HVH*, pp. 39-45; Patterson, *Nuclear Power*, pp. 99ff.
376. *HVH*, p. 46.
377. Ibid.
378. Ibid., p. 48.

379. Ibid.

380. For projected death rates from these tailings, see Dr Lochstel's calculations in *HVH*, pp. 49-51; see also Patterson, op.cit., pp. 91-5.

381. The story of Joe Harding has been taken from Taylor G. Moore III, 'Joe Harding's Death List: The Growing Toll of Workers in the Uranium Enrichment Industry', *The Progressive*, January 1980, pp. 24-9.

382. *HVH*, p. 54.

383. Ibid, pp. 59, 60.

384. Ibid.

385. Ibid., pp. 66-71.

386. Ibid., p. 69.

387. Connecticut Health Department Registration Reports, US Monthly Vital Statistics Reports, 1970-1975.

388. *HVH*, p. 71.

389. Dick Brukenfeld, 'A New German Study Challenges the NRC's Assurances', *Washington Post*, 11 November 1971, p. B1.

390. Ibid.

391. *The Guardian*, 15 May 1979, p. 3.

392. Jungk, *Nuclear State*, p. 14.

393. Ibid., p. 15.

394. Ibid., p.17.

395. Ibid., p. 18.

396. *HVH*, p. 100.

397. Ibid., p. 104.

398. *The Guardian*, 6 June 1979, p. 1.

399. *HVH*, p. 103.

400. Berger, *Nuclear Power : The Unviable Option*, p. 106.

401. Ibid., p. 107.

402. Ibid., p. 109.

403. Ibid., pp. 113f.; *HVH*, 122-24.

404. *HVH*, pp. 123, 124.

405. Berger., op.cit., p. 117; also Patterson, op.cit., pp. 103-14.

406. In Jungk, *Nuclear State*, pp. 118, 119.

407. Nuclear Regulatory Commission Contract No. AT(49–24)–0190, 31 October 1975, p. 1, (emphasis mine).

408. *Nuclear State*, p. 133; for details on Flood's and Grove-White's findings, see their pamphlet, 'Nuclear Prospects', London 1976.

409. *Nuclear State*, p. 132.

410. The following is taken from my article 'The Mysterious Case of Karen Silkwood' *The Ecologist*, Vol. 9, No. 8, November/December 1979.

411. *International Herald Tribune*, 19 May 1979, p. 1.

412. Ibid.

413. In *Nuclear State*, p. 142.

414. Ibid., pp. 145, 146.

415. Ibid., p.4.

416. *The Observer*, 1 January 1978.

417. In Lifton 'Nuclear Energy and the Wisdom of the Body', *Bulletin*

of the Atomic Scientists, Vol. 34, No. 7, September 1976, p. 18.

418. B. Flowers, 'Nuclear Power: A Perspective of the Risks, Benefits and Options', *Bulletin of the Atomic Scientists*, Vol. 34, No. 3, March 1978, p. 22

419. A. Lovins, *Is Nuclear Power Necessary?*, London 1979, p. 9.

420. Ibid., p. 11.

421. Ibid., pp. 3-5.

422. Gerald Leach et al., *A Low Energy Strategy for the United Kingdom*, London 1979, passim.

423. In a letter to the *New York Times*, 17 October 1975.

424. Berger, *Nuclear Power*, p. 231.

425. Lovins, *Is Nuclear Power Necessary?*, p. 19.

426. Ibid, p. 24; see also Lovins, *Soft Energy Paths: Toward a Durable Peace*, London 1977; New York 1979, passim.

427. Berger, op.cit., p. 257.

428. Philip Steadman, *Energy, Environment and Building*, Cambridge, England 1975.

429. W. Steifert et al., (eds.), *Energy and Development: A Case Study*, Cambridge, Mass. 1973.

430. In Berger, op.cit., p. 259.

431. Lovins, *Is Nuclear Power Necessary?*, p. 25.

432. Ibid.

433. Ibid., p. 26.

434. V. Taylor, *Energy: The Easy Path*, Los Angeles 1979; see also Lovins, *Is Nuclear Power Necessary?*, pp. 17, 18.

435. Ibid., p. 22.

436. L.L. Whyte, *The Next Development in Man*, New York 1948, p. 122.

437. Ibid.

438. Ananda Coomaraswamy, *Am I My Brother's Keeper?*, New York 1964, p. 3.

439. Ibid., p. 64.

440. Thomas Merton, *Gandhi On Non-Violence*, New York 1964, p. 3.

441. Alvin Toffler, *Future Shock*, London 1970, p. 4.

442. Ibid., p. 5.

443. Ibid., p. 15.

444. Hans Küng, *On Being A Christian*, London 1977, p. 34.

445. In Toffler, op.cit., p. 15.

446. J. Lifton, *History and Human Survival*, New York 1961, p. 157.

447. In A. Koyré, *Discovering Plato*, New York 1945, p. 108.

448. Henry Wieman, *The Source of Human Good*, New York 1946, pp. 40, 41.

449. In Merton, op.cit., p. 9.

450. Ibid., p. 11.

451. Hannah Arendt, *The Human Condition*, Chicago, 1958, p. 240.

452. In Merton, op.cit., pp. 13, 14.

453. Ibid., p. 43.

454. Ibid., p. 50.

455. Ibid., p. 33.

456. John Rickman, (ed.), *Civilization, War and Death: A Selection of the Writings of Freud*, London 1953, p. 82.

457. Bigelow, *The Dawn Warriors*, p. 9.

458. Ardrey, *African Genesis*, p. 325.

459. Ibid.

460. In Merton, op.cit., p. 32.

ACKNOWLEDGMENTS

The tables on pages 116, 117 and 169 and referred to in notes 308, 309 and 364 are reprinted by permission of *Bulletin of the Atomic Scientists*, a magazine of science and public affairs. Copyright © 1976 and 1978 by the Educational Foundation for Nuclear Science, Chicago, Illinois.

The diagrams on pages 138-39 are reproduced by permission of Penguin Books Ltd from W.C. Patterson, *Nuclear Power*, 1976.

INDEX